S0-DJQ-131

CHICAGO PUBLIC LIBRARY
HAROLD WASHINGTON LIBRARY CENTER

R0034175780

TJ   Hoffman, Edward G.
1166
.H58   Basic shop
1983   measurement

BUSINESS/SCIENCE/TECHNOLOGY DIVISION

DATE			

Business/Science/Technology
Division

© THE BAKER & TAYLOR CO.

BASIC SHOP MEASUREMENT

NMTBA Shop Practices Series

BASIC SHOP MEASUREMENT
BLUEPRINT READING
SHOP MATH
SHOP THEORY

1 millimeter (mm) = 0.03937 inch (in.)   1 inch (in.) = 25.4 millimeters (mm)

mm	*in.*	*in.*	*mm*
1.0 mm	= 0.03937 in.	1.0 in.	= 25.4 mm
0.1 mm	= 0.003937 in.	0.1 in.	= 2.54 mm
0.01 mm	= 0.0003937 in.	0.01 in.	= 0.254 mm
0.001 mm	= 0.00003937 in.	0.001 in.	= 0.0254 mm
0.0001 mm	= 0.000003937 in.	0.0001 in.	= 0.00254 mm

millimeter value = inch value × 25.4   inch value = millimeter value × 0.03937

Figure 1-6 Converting between SI and English measurements.

CONVERTING MEASUREMENTS

Converting measurements between the SI and English systems is an easy task. All you need to remember are the conversion values. These values, shown in Figure 1-6, are exact and the arithmetic is simple.

SELF TEST

1. List the three principal reasons for measurement.
2. For what two purposes are measurements made in the machine shop?
3. What was the first known reference standard?
4. What four other standards evolved from this first standard?
5. What was the first worldwide measurement system called?
6. What was the purpose of the Metric Conversion Act of 1975?
7. What unit is the base unit for the English system? For the SI?
8. What units of measurement are generally used for manufacturing purposes in both systems?
9. What is the conversion value for measurements from English to SI?
10. What is the conversion value for measurements from SI to English?

Answers to Self Test

1. a. To control the way we make parts
   b. To control the way others make parts for us
   c. To provide a graphic description of the part
2. a. To find an exact size
   b. To inspect finished parts
3. Egyptian royal cubit
4. a. The Greek Olympic cubit
   b. The Greek Olympic foot
   c. The Roman foot
   d. The Anglo-Saxon foot
5. The metric system
6. Coordinating the increased use of the metric system in the United States
7. The international yard. The meter
8. The international inch and the millimeter
9. Inch value × 25.4 = millimeter value
10. Millimeter value × 0.03937 = inch value

UNIT 2

The Language of Measurement

In addition to numbers, there are also many words and terms used to describe measurements. These words and terms form the language of measurement. The language of measurement is not nearly as exact as the numbers it describes. Terms are sometimes confused, used improperly, or even ignored in the daily routine of shop work, but sometimes you'll want to be precise in what you mean, so it's best to know the terms and how each is properly used.

MEASUREMENT FACTORS

When measuring, several factors must work together to insure a proper measurement. With any measurement you must have a *standard*, a *reference point*, a *measured point*, and a *line of measurement*.

STANDARD
A standard is an established and known value that is used to measure an unknown quantity. In linear measurement, the international standards are the inch and the millimeter.

Typical standards used in the shop include the rule and the micrometer (Figure 2-1). These standards, in addition to showing full inch and millimeter values, are divided into smaller units. These smaller units are either fractional or decimal parts of the whole inch or millimeter.

REFERENCE POINT
The reference point in measurement is the base from which the measurement is taken. It can be the anvil of a micrometer, the solid jaw of a vernier caliper, or the end or any point on a rule (Figure 2-2). It is the point of origin for any measurement or, put another way, it is the point where a measurement begins.

MEASURED POINT
The measured point is the line or edge that is measured. It is the end opposite the reference point, as shown in Figure 2-2. This is the terminal point of the measurement. In other words, it is the end of the measurement.

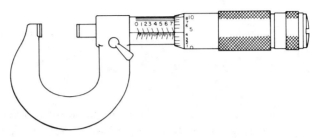

Figure 2-1 Typical standards used for measuring in the shop.

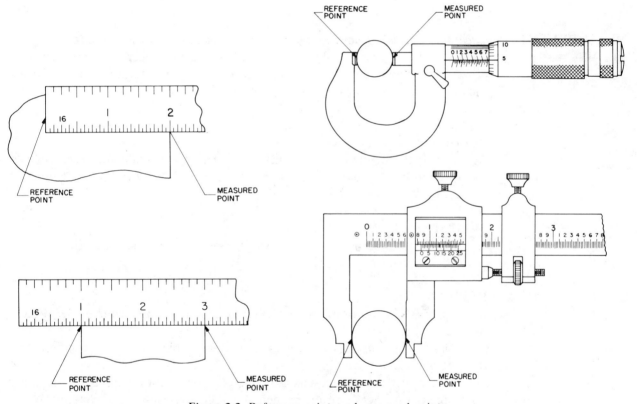

Figure 2-2 Reference points and measured points.

LINE OF MEASUREMENT

The line of measurement (Figure 2-3) is an imaginary straight line between the reference point and the measured point, parallel to the axis of the measuring tool. The position of this line in relation to the part determines the exactness of the measurement. If the line of measurement is tilted, the measurement will be in error, resulting in "parallax error." The measuring tool must always be held so that the line of measurement is perpendicular to the reference point

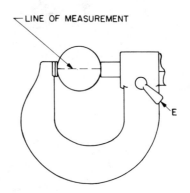

Figure 2-3 The line of measurement.

and the measured point. Stated simply, if you do not make a straight measurement, your reading will be inaccurate and will indicate that you have more stock than you actually have.

MEASUREMENT TERMS

Accuracy, precision, and *reliability* are terms used to describe measurements. Each of these terms describes different elements or concepts in the measuring process. To understand the measuring process, you must know the meaning of each of these terms.

ACCURACY
Accuracy is the degree of conformity to an established standard. Accuracy can also be considered as a comparison of the desired results with the actual results.

PRECISION
Precision is the exactness of the measuring process and its repeatability.

RELIABILITY
Reliability is a condition where the actual results are the same as the desired or predicted results.

To clarify these definitions, let us look at an often used example to see how these elements are related.

Assume that three individuals are competing in a shooting match. If each has 10 shots per turn, let us compare their accuracy, precision, and reliability.

For the purpose of this description we can define accuracy as the comparison of the shots inside the target to those outside the target. Precision is related to the grouping of the shots in a pattern. Reliability is the consistency of the shooting from one round to the next.

In the first round of shooting the targets appear as shown in Figure 2-4. From the results we can say the following about each shooter.

Al, with a score of five, is more accurate than Barb, who scored zero. However, since Barb's pattern is closer together than Al's, we can say Barb is more precise. Charlie has a pattern that is more precise than Al's, but not as precise

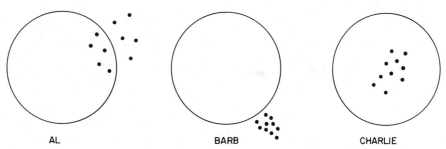

Figure 2-4 Relationship of accuracy and precision.

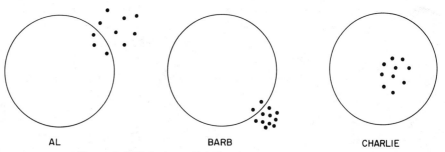

Figure 2-5 Relating reliability to accuracy and precision.

as Barb's. However, since Charlie scored a perfect ten, he is more accurate than either Al or Barb.

Now, if we change the distance from the target, we can find the reliability of the shooters. Assuming we move the shooters back another 50 ft, their targets would appear as shown in Figure 2-5. We can now say the following about each.

Al, with a score of three, is not as accurate this time, but he is still more accurate than Barb, who only scored two. Barb, however, is still more precise than Al. Charlie held the same pattern as before—still not as precise as Barb, but more precise than Al. Since Charlie scored another ten, he is still more accurate than Al or Barb.

In this round both Al and Barb were not reliable in their shooting because there was a change in their score from the first round. Charlie, however, is reliable because he matched his first-round score.

As a final test to see who is the best shot, we can reduce the size of the target (Figure 2-6). Now let us see how the shooters compare.

In this last round Al has a score of one. He is still more accurate than Barb, who missed completely. Barb, however, is still more precise than either Al or Charlie, since her pattern is the closest. Charlie has maintained his 100% record

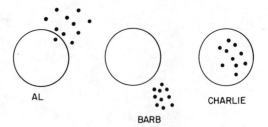

Figure 2-6 Relationship of reliability, accuracy, and precision.

and is still more accurate than Al or Barb. Since Charlie scored a perfect ten-ten-ten, he is also the most reliable of the shooters.

From this we can summarize by saying that Charlie is the most reliable and the most accurate, Barb, while the least accurate, is the most precise, and Al, even though more accurate than Barb, is the least precise and least reliable of the shooters.

In this example we compared people target shooting. We could also have compared the rifles simply by using one person to shoot all the rifles. Likewise, in measurement both the measuring tool and the machinist must be considered as factors when accuracy, precision, and reliability are involved. In any case these terms are all related, yet very different.

The accuracy of a measuring tool is related to how close the measurement comes to the actual size. The precision of a measuring instrument is determined by how closely it can repeat identical measurements. The reliability of a measuring tool is decided by how consistently it can obtain the desired or predicted results.

FACTORS THAT AFFECT MEASUREMENT

Perfection in measurement or manufacturing is impossible. There will always be slight variations in every part. Designers realize this and allow for small variations in manufacture. To control the variations and to insure that each part will function as intended, the designers specify *limits,* or *tolerances,* on important *dimensions.*

The terms dimension, tolerance, and limits describe factors that directly affect and influence measurement.

DIMENSIONS

A dimension is the exact size of the finished part. As shown in Figure 2-7, the dimension is 2.000 in. and indicates the base size of the part.

TOLERANCE

The tolerance is the total amount of permitted variation from the base size. As shown in Figure 2-7, the tolerance is plus or minus 0.005 in., or 0.010 in. total.

LIMITS

The limits are the maximum and minimum size of the completed part as described by the tolerance. In the case shown in Figure 2-7, the limits are 1.995 in. for the smallest size and 2.005 in. for the largest size. Any part that falls within these sizes is acceptable.

Limits set the permissible sizes for a part and must be observed. Trying to maintain a tolerance closer than that specified is both wasteful and inefficient. You should always work toward producing parts that fall within the limits. There is no reason to try to hold a tolerance of ±0.0001 in. on a part that has a tolerance value of ±0.030 in. All you will do trying is spend extra time and slow your production.

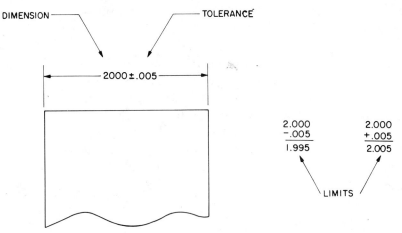

Figure 2-7 Dimensions, tolerances, and limits.

SELF TEST

1. Identify the parts labeled 1 to 4 in Figure 2-8.

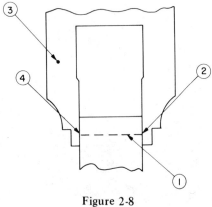

Figure 2-8

2. Define the following terms.
    a. Accuracy.
    b. Precision.
    c. Reliability.

3. Answer the following questions concerning Figure 2-9.
    a. What is the dimension?
    b. What is the tolerance?
    c. What are the limits?

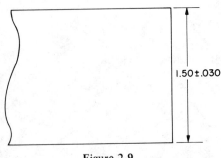

Figure 2-9

Answers to Self Test

1. a. Part 1 is the line of measurement
   b. Part 2 is the measured point
   c. Part 3 is the standard
   d. Part 4 is the reference point
2. a. Accuracy is the degree of conformity to an established standard; it is also the comparison of the desired results to the undesired results
   b. Precision is the repeatability of the measuring process
   c. Reliability is a condition where the actual results are the same as the desired or predicted results
3. a. 1.50 in.
   b. ±0.030 in.
   c. 1.470 to 1.530 in.

SECTION II

GRADUATED MEASURING TOOLS

UNIT 3 — **Steel Rules**

UNIT 4 — **Adjustable Squares and Rule Attachments**

UNIT 3

Steel Rules

The simplest and most versatile measuring tool a machinist uses is the steel rule. Steel rules are thin, steel blades with precisely etched graduations. They are available in a wide variety of styles and sizes to suit their many applications. Learning to read steel rules and care for them properly is one of the first steps to becoming a skilled machinist.

TYPES OF STEEL RULES

Several kinds of steel rules are used for many different measuring tasks commonly performed in the machine shop.

SPRING TEMPERED RULES

The spring tempered rule (Figure 3-1) is the most common type of rule used for general shop measurement. The most popular sizes are 6 and 12 in., although these rules are made in lengths ranging from 1 to 144 in. Spring tempered, or rigid, rules are well suited for many measuring tasks. When space is limited, however, or when a rule must bend around a part to make a measurement, a narrow rule or flexible rule should be used.

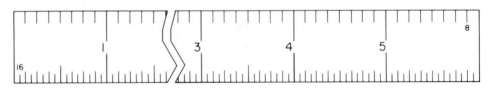

Figure 3-1 Spring tempered rule.

NARROW RULES

The narrow rule (Figure 3-2) is used to make measurements in places where a standard, rigid rule cannot fit. These rules are generally made in widths from $\frac{3}{16}$ to $\frac{1}{4}$ in. and lengths of 6 and 12 in.

Figure 3-2 Narrow rule.

Figure 3-3 Flexible rule.

FLEXIBLE RULES

The flexible rule (Figure 3-3) is used in places where the rule must be slightly bent to make a measurement. These rules are thinner and narrower than the standard rigid rule. They can often be used in places where a rigid rule will not fit. Flexible rules are also made in lengths of 6 and 12 in.

HOOK RULES

The hook rule (Figure 3-4) is made with either a fixed or adjustable hook on one end. This hook locates the reference point of the rule and permits accurate measurements, even in places where the end of the rule cannot be seen (Figure 3-5). Hook rules are made in either narrow or standard widths and in lengths of 6 and 12 in.

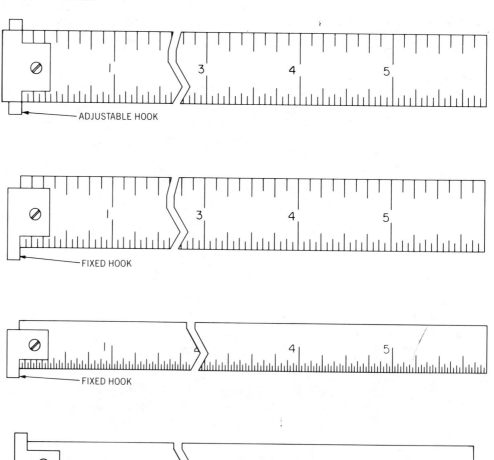

Figure 3-4 Hook rules.

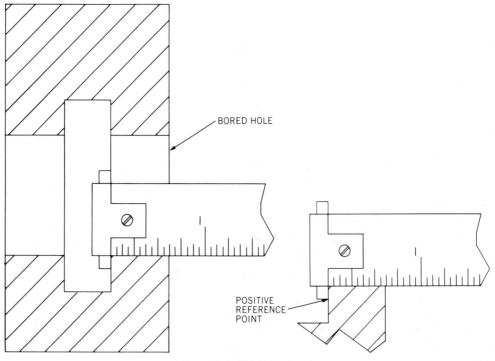

Figure 3-5 Using hook rules.

RULE SETS

A rule set (Figure 3-6) consists of a handle and five small rules: $\frac{1}{4}$, $\frac{3}{8}$, $\frac{1}{2}$, $\frac{3}{4}$, and 1 in. This set is useful for measuring in limited spaces such as slots, grooves, and keyways, where other rules cannot be used.

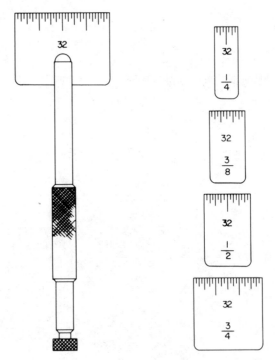

Figure 3-6 Rule set.

SECTION II GRADUATED MEASURING TOOLS

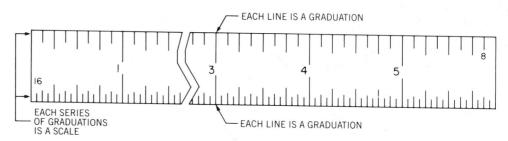

Figure 3-7 Scales and graduations.

READING THE STEEL RULE

As you look at a steel rule, notice the series of equally spaced lines along each edge. These lines are called *graduations* and determine the *scale* of the rule. The scale of a rule is the relationship between the number of graduations and the unit length, usually 1 in. or 1 mm. In other words, each series of lines, or graduations, on each edge of a rule is a scale (Figure 3-7).

Most rules used in the shop have scales that divide inches into fractional parts. Other scales used for rules include decimal inch and millimeter scales (Figure 3-8). Almost any combination of these scales is available. Figure 3-9 shows the most popular scale combinations.

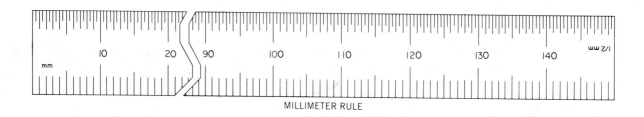

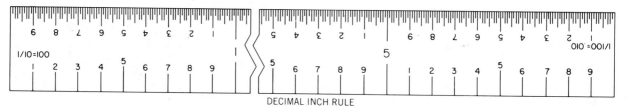

Figure 3-8 Decimal and millimeter rules.

GRADUATION NUMBER	FIRST EDGE	SECOND EDGE	THIRD EDGE	FOURTH EDGE
3	32nds	64ths	50ths	10ths
4	64ths	32nds	16ths	8ths
5	10ths	100ths	64ths	32nds
10	64ths	32nds	—	—
12	100ths	50ths	—	—

Figure 3-9 Common scale combinations.

Unit 3 Steel Rules   21

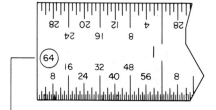
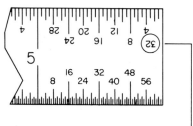

Figure 3-10 Discrimination number indicates number of divisions per inch.

When reading a rule, first identify the scale. Most rules have a small number at the beginning of each scale that indicates the number of divisions per inch (Figure 3-10). This number also indicates the *discrimination* of the scale. Discrimination refers to the finest division of a measuring tool that may be reliably read. For example, the rule shown in Figure 3-11 has two scales, marked 8 and 16. The finest division on the top scale is 8, or $\frac{1}{8}$ in. The discrimination of this scale is also $\frac{1}{8}$ in. The finest division on the bottom scale is 16, or $\frac{1}{16}$ in. The discrimination of this scale is $\frac{1}{16}$ in.

On some rules you will also notice a series of numbers above the graduation marks (Figure 3-12). These numbers are called quick reading references and help you read the rule quickly and accurately. Another method commonly used to help define the graduations is the staggering of graduation lines. Since the lengths of these lines vary, you can determine an exact measurement more quickly than if you had to count each line (Figure 3-13).

Fractional rules divide each inch unit into 8, 16, 32, or 64 equal parts. These scales are always read as fractional parts of one inch. For example, each line on the 8 scale is equal to $\frac{1}{8}$ in., each line on the 16 scale is equal to $\frac{1}{16}$ in., and so

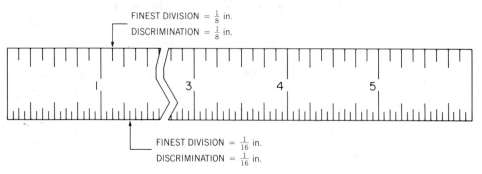

Figure 3-11 Discrimination of scale.

Figure 3-12 Quick reading reference numbers.

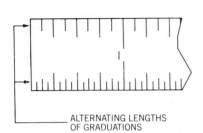
Figure 3-13 Staggering graduation for easier reading.

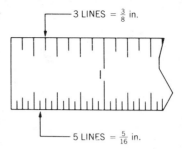
Figure 3-14 Reading a fractional rule.

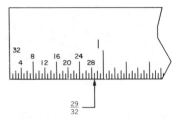
Figure 3-15 Using the reference numbers to read a measurement quickly.

on. In practice, a rule is read by counting the number of graduations between the reference point and the measured point. This reading then serves as the top number or numerator of a fraction. The scale number acts as the denominator, the lower number, of the fraction. For example, if you read 3 lines on the 8 scale, the reading is $\frac{3}{8}$ in. Likewise, if you read 5 lines on the 16 scale, the measured value would be $\frac{5}{16}$ in. (Figure 3-14). In cases where your reading is a reducible fraction, such as 12 lines on the 16 scale, or $\frac{12}{16}$ in., you will have to reduce the fraction to its lowest terms: $\frac{12}{16}$ in. $\times \frac{4}{4} = \frac{3}{4} \frac{\cancel{12}}{\cancel{16}} \times \frac{\cancel{4}}{\cancel{4}} \frac{1}{1} = \frac{3}{4}$ in. This measurement should be expressed as $\frac{3}{4}$ in., not $\frac{12}{16}$ in.

On the $\frac{1}{32}$ and $\frac{1}{64}$ scales, you will notice the quick reading reference numbers are above each $\frac{1}{8}$-in. graduation. These numbers are used as markers to indicate the number of divisions at each $\frac{1}{8}$-in. division. Since the $\frac{1}{64}$ scale has twice as many graduations as the 32 scale, the reference numbers on the 64 scale are double those on the 32 scale. Look at the rule shown in Figure 3-15. The reading shown is $\frac{29}{32}$ in. To read it, simply find the closest reference number and either add or subtract the lines from the measured point to the reference number. In this case, $\frac{29}{32}$ is one line past the quick reading reference number 28 so 28 + 1 = 29. Since the scale you used was 32, the reading is $\frac{29}{32}$ in. Another point to remember when reading a rule is to add the whole inch units, as shown in Figure 3-16. This reading is $1\frac{7}{8}$ in., not just $\frac{7}{8}$ in.

Reading a decimal inch scale is also simple. Again, the first step is to identify the scale. Decimal inch scales are generally graduated in 10ths, 50ths, and 100ths of an inch. This style of graduation is sometimes preferred over fractional scales. Since the readings can be directly transferred into decimal units, you do not have to refer to a decimal equivalent chart.

When using a decimal inch rule, note the position of the measured point and count the number of graduations from the reference marker to the measured point. In the reading shown in Figure 3-17, the measurement is 80 + 7, or 87.

Figure 3-16 Always count whole inch units.

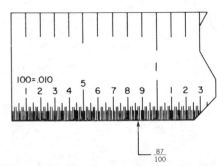
Figure 3-17 Reading a decimal inch rule.

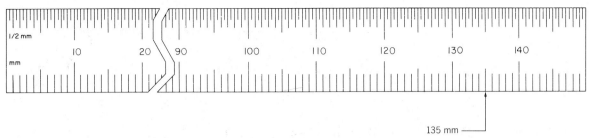

Figure 3-18 Reading a millimeter rule.

The scale is 100, so the measurement is $\frac{87}{100}$ in. The reference numbers used on this scale indicate the number of divisions in groups of 10. So the number 1 equals 10, 2 equals 20, 3 equals 30, and so on for each inch. To convert this reading to a decimal value, simply multiply both the numerator and denominator by 10.

$$\frac{87}{100} \times \frac{10}{10} = \frac{870}{1000} \quad \text{or} \quad 0.870$$

Each graduation line on the $\frac{1}{100}$ scale is equal to 0.010 in. Each line on the $\frac{1}{50}$ scale is equal to 0.020 in., and each line on the $\frac{1}{10}$ scale is equal to 0.100 in.

Millimeter rules are graduated in millimeters and half millimeters. These are read in millimeter units, not centimeters or decimeters. The System International (SI) unit used for basic machine shop measurements is the millimeter. As shown in Figure 3-18, the dimension is read as 135 millimeters, not 13.5 centimeters.

Now that you know how to read a rule, let's try a few practice problems. Read the rules in Figure 3-19 to the values indicated.

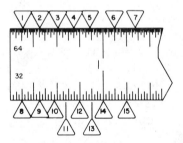

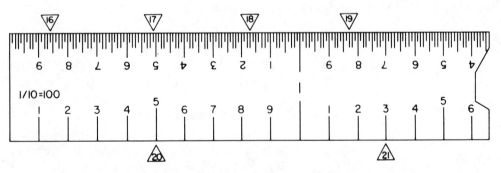

Figure 3-19 Practice problems in reading a rule.

24 SECTION II GRADUATED MEASURING TOOLS

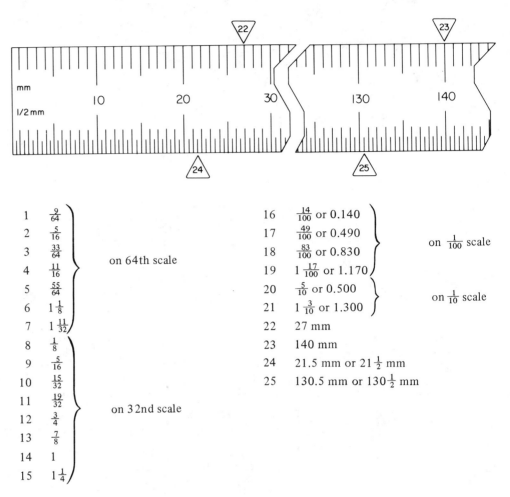

1	$\frac{9}{64}$	
2	$\frac{5}{16}$	
3	$\frac{33}{64}$	
4	$\frac{11}{16}$	on 64th scale
5	$\frac{55}{64}$	
6	$1\frac{1}{8}$	
7	$1\frac{11}{32}$	
8	$\frac{1}{8}$	
9	$\frac{5}{16}$	
10	$\frac{15}{32}$	
11	$\frac{19}{32}$	on 32nd scale
12	$\frac{3}{4}$	
13	$\frac{7}{8}$	
14	1	
15	$1\frac{1}{4}$	

16	$\frac{14}{100}$ or 0.140	
17	$\frac{49}{100}$ or 0.490	on $\frac{1}{100}$ scale
18	$\frac{83}{100}$ or 0.830	
19	$1\frac{17}{100}$ or 1.170	
20	$\frac{5}{10}$ or 0.500	on $\frac{1}{10}$ scale
21	$1\frac{3}{10}$ or 1.300	
22	27 mm	
23	140 mm	
24	21.5 mm or $21\frac{1}{2}$ mm	
25	130.5 mm or $130\frac{1}{2}$ mm	

Figure 3-19 (*continued*)

MAKING DIRECT MEASUREMENTS WITH A STEEL RULE

Measurements made with a steel rule can be grouped into two general classes: direct and indirect. Direct measurements are made with a rule alone. Indirect, or transfer, measurements are made with measuring tools such as inside and outside calipers, which are not read directly but are checked against a rule.

When making any direct measurement, consider instrument error, manipulation error, observational error, and error of bias.

INSTRUMENT ERROR

Instrument error is primarily caused by poor quality, wear, or abuse. Make sure your rule is in good condition and accurately made. If you were measuring to $\frac{1}{64}$ or $\frac{1}{100}$ of an inch, it is doubtful that you could get an accurate reading from a wooden ruler. Make sure your tools are accurate enough to do the job. Never try to measure to an accuracy finer than the discrimination of your rule. In other words, do not try to read 64ths on a 32nds scale. Always use the proper scale. Next, see that your rule is in good condition. Remove any burrs with an oilstone. Carefully clean and check the rule before using it. Never abuse a rule. Do not use your rule for anything but measuring.

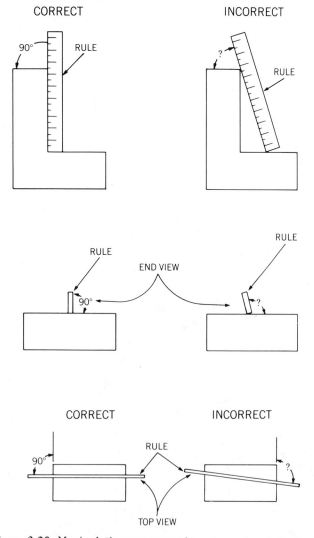

Figure 3-20 Manipulative error must be corrected and eliminated.

MANIPULATIVE ERROR

Manipulative error is caused by improper handling of the rule during a measurement. When measuring, hold your rule squarely to the workpiece (Figure 3-20). Any tilt or misalignment will affect the measurement.

When beginning a measurement, know where the reference point is (Figure 3-21). Any misalignment can affect the measurement. Figure 3-22 shows several correct and incorrect measuring conditions. When you use a rule, always make sure you use it properly.

OBSERVATIONAL ERROR

Observational, or parallax, error is caused by misalignment between the rule and the observer. To read a rule accurately, hold the rule properly and make your reading from the correct viewing angle. Figure 3-23 shows how parallax error can affect your measurement.

SECTION II GRADUATED MEASURING TOOLS

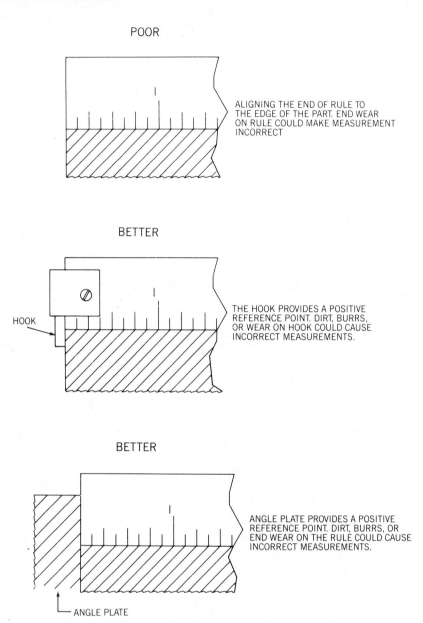

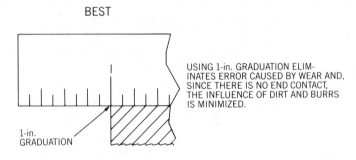

Figure 3-21 Locating the reference point of the rule.

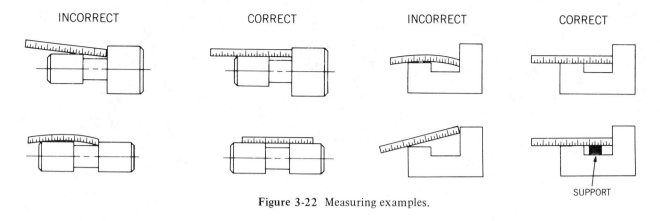

Figure 3-22 Measuring examples.

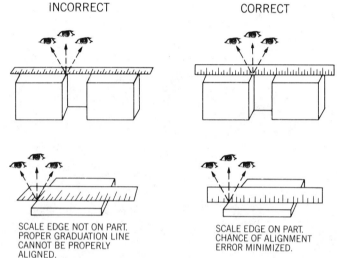

Figure 3-23 Parallax error.

ERROR OF BIAS
The error of bias is an unconscious influence on a measurement. All machine operators want to make a good part, so unconsciously you may try to push or pull the rule to make it indicate the measurement you want. You must always be careful to make every measurement as accurately as possible.

MAKING INDIRECT MEASUREMENTS WITH A STEEL RULE

Indirect, or transfer, measurements are made by first checking the size of a workpiece with either an inside or outside caliper and then transferring the setting to a rule where the size can be measured (Figure 3-24). Transfer measurements must be made carefully due to the greater chance of error.

ERRORS IN INDIRECT MEASUREMENT
Possibly the two greatest problems when making transfer measurements arise from alignment of the contacts and "feel." To measure accurately any part with a caliper-style instrument, align the contact points across the axis of the

28 SECTION II GRADUATED MEASURING TOOLS

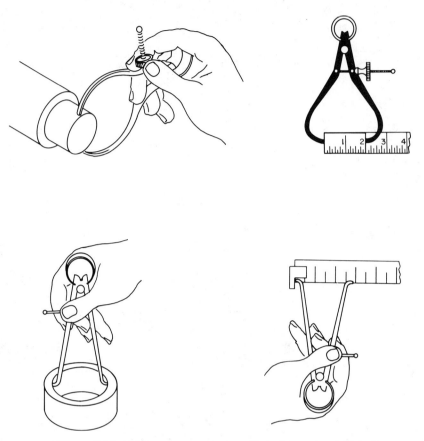

Figure 3-24 Making measurements with inside and outside calipers.

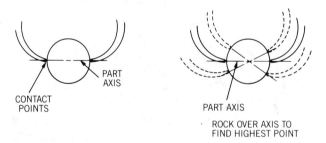

Figure 3-25 Locating part axis with calipers.

part, as shown in Figure 3-25. If one contact point is higher or lower than the other point, an inaccurate measurement will result.

"Feel" is the resistance between the contact points and the workpiece. This resistance is in the form of a slight drag when the caliper is moved on the workpiece. Feel is very important to accurate transfer measurements. The lighter the feel, the better. A light feel allows you to make much more accurate measurements and helps to insure the proper alignment of contact points. Figure 3-26 shows several correct and incorrect methods of using calipers to make transfer measurements.

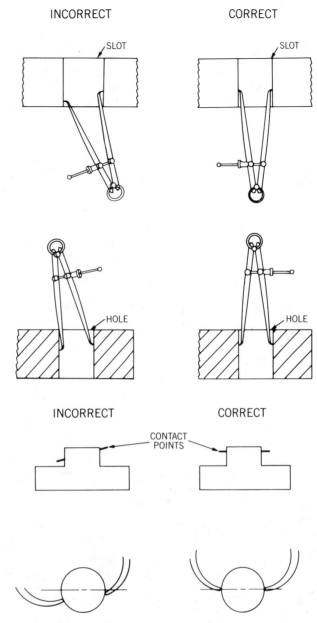

Figure 3-26 Proper use of calipers is important to accuracy.

CARE OF STEEL RULES

A steel rule is a precision instrument. With proper use and care, it will serve you well. The following suggestions are offered to help you make the best use of your steel rules.

1. Use the proper rule for the job. You should not use a 12-in. rule to measure a 2-in. part.

2. Select the proper scale. Do not try to read a rule any closer than the discrimination of the scale used.

3. Treat a rule carefully. Do not leave it on a machine or bench while you are working. Put it in your apron pocket or toolbox.

4. Keep your rule clean and lightly oiled when not in use. Frequently wipe your rule with oil to prevent rust and stains.

5. Keep your rule away from moving cutters and moving workpieces.

SELF TEST

1. What are the most common graduations found on steel rules used in the machine shop? (Circle one.)

   a. 12ths, 16ths, 32nds, and 64ths

   c. 64ths, 32nds, 16ths, and 8ths

   c. $\frac{1}{2}$ mm and 1 mm

   d. 10ths, 50ths, and 100ths

2. What is each series of equally spaced lines called? (Circle one.)

   a. Discrimination

   b. Parallax

   c. Graduations

   d. Scales

3. What is the finest division of a measuring tool that can be read reliably called? (Circle one.)

   a. Graduation

   b. Discrimination

   c. Accuracy

   d. Scale

4. List five different types of steel rules.

5. What are the three most common types of scales?

6. List the four primary types of errors in direct measurements.

7. What two sources of error must be considered in transfer measurement?

8. What are two common transfer measuring tools?

9. List five ways to use and care for a steel rule.

Answers to Self Test

1. b.
2. d.
3. b.
4. a. Spring tempered rule
   b. Narrow rule
   c. Flexible rule
   d. Hook rule
   e. Rule set
5. a. Fractional inch
   b. Decimal inch
   c. Millimeter

6. a. Instrument error
   b. Manipulation error
   c. Observation error
   d. Error of bias
7. a. Position of contact points
   b. Feel
8. a. Inside calipers
   b. Outside calipers
9. a. Use the proper rule for the job
   b. Select the proper scale
   c. Treat a rule carefully
   d. Keep your rule clean and oiled
   e. Keep your rule away from moving cutters or workpieces

UNIT 4

Adjustable Squares and Rule Attachments

The steel rule (Figure 4-1) is a basic measuring tool. Other tools and attachments have been developed from it and for it. The combination square, double square, diemaker's square, and depth gage are all tools that use a steel rule for measuring. In addition, attachments such as keyseat clamps and rule clamps add more versatility to the basic steel rule.

ADJUSTABLE SQUARES

Adjustable squares use the basic steel rule in combination with movable heads to check squareness and to make measurements. The principal types of adjustable squares in common use today are the combination square, the double square, and the diemakers' square.

COMBINATION SQUARES

The combination square set consists of a steel rule, square head and center head. A protractor head, also used with this set, is discussed in Unit 11.

As shown in Figure 4-2, the combination square is a very versatile tool that can be used for checking squareness, measuring, checking 45° angles, drawing lines parallel to an edge, and checking level. With a center head, it can also be used to locate and mark the center of cylindrical workpieces (Figure 4-3).

The steel rule used in a combination square is grooved on one side to allow the square head and center head to be positioned anywhere along the length

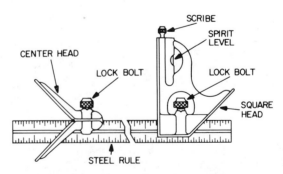

Figure 4-1 The combination square.

32

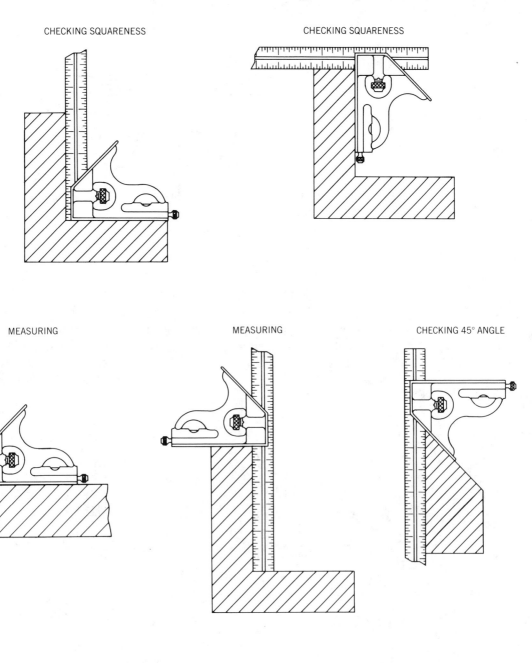
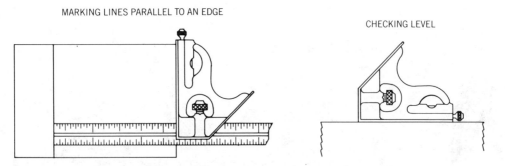

Figure 4-2 Using the combination square.

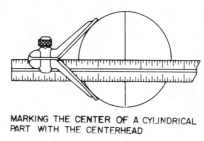

MARKING THE CENTER OF A CYLINDRICAL
PART WITH THE CENTERHEAD

Figure 4-3 Using the center head.

of the rule. The most common sizes of combination squares are 6 and 12 in. The rules used with a combination square are available in lengths of 4 and 6 in. for the 6-in. set and 9, 12, 18, and 24 in. for the 12-in. set. Figure 4-4 shows several conditions you must watch for when using a combination square.

DOUBLE SQUARES

Double squares (Figure 4-5) are a variation of the combination square and are useful for checking squareness, measuring, and drawing lines parallel to an edge. Double squares are generally smaller than combination squares and can be used where a combination square would be too large. The rule used with a double square is also grooved to allow the square head to be set at any point along the

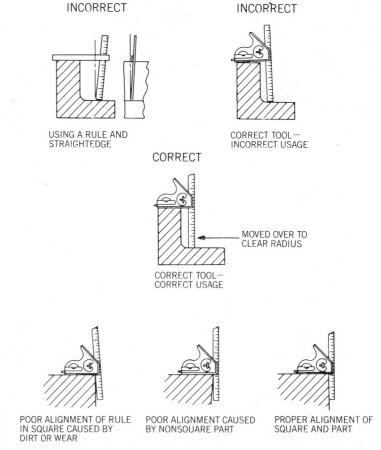

Figure 4-4 Methods of making measurements with the combination square.

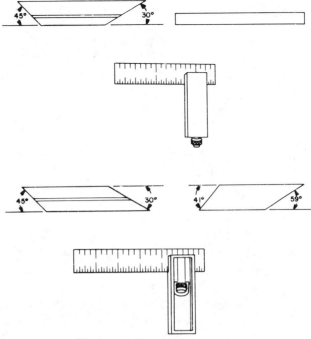

Figure 4-5 The double square.

length of the rule. Double squares are available in $2\frac{1}{2}$, 4, and 6-in. sizes. In addition to the standard steel rule, many double squares also have blades with an angle at each end to check 45 and 30° angles, as well as a blade for checking drill points.

DIEMAKERS' SQUARES

Diemakers' squares (Figure 4-6) are variations of the double square, the main difference being the tilt blade feature of the diemakers' square. On the square shown at the left of Figure 4-6, the blade is tilted to measure any angle up to 10°. This is useful in checking parts with slight angles, such as die clearances.

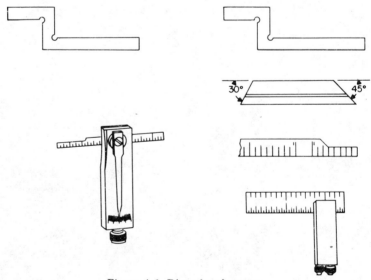

Figure 4-6 Diemakers' square.

On the type shown on the right of the figure, the angle is set by tightening the small knurled screw until the desired angle is reached. The square is then locked in this position by tightening the large knurled screw. The actual reading is then obtained by comparing the setting to an angle-measuring tool, such as a protractor. This type of square also has a variety of blades to suit its many applications.

VARIATIONS OF THE ADJUSTABLE SQUARE

The depth gage and the slide caliper, variations of the adjustable square, have been developed to meet the wide variety of machine shop measuring requirements.

DEPTH GAGE

The depth gage (Figure 4-7) is used for measuring the depth of holes, shoulders, and slots. The basic construction of this tool includes a narrow rule and a movable head. In use the rule is first located in the part detail to be measured. Loosen the lock nut and lower the rule to the bottom of the hole or slot. Once the rule touches bottom, tighten the lock nut and remove the depth gage. Take the measurement directly from the bottom edge of the head (Figure 4-8). A few common mistakes in using the depth gage are shown in Figure 4-9.

SLIDE CALIPERS

Slide calipers (Figure 4-10) are a direct variation of the steel rule. With this measuring tool, both the reference point and the measured point are located by the jaws on the caliper. The measurement is taken from the sliding rule at the positions marked IN or OUT. The reference lines are positioned this way

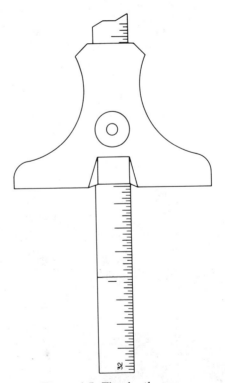

Figure 4-7 The depth gage.

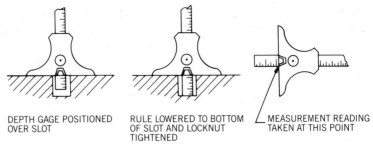

Figure 4-8 Making measurements with the depth gage.

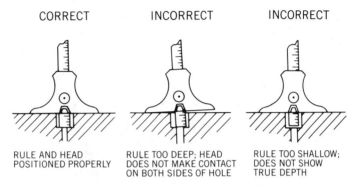

Figure 4-9 Problems in using the depth gage.

to allow for the thickness of the measuring tips on the ends of the jaws. These tips, or nibs, are usually $\frac{1}{4}$ in. wide when the jaws are closed. Always make sure you are using the proper reference line for the measurement you are making (Figure 4-11). Slide calipers are generally available in sizes of 3, 5, and 6 in.

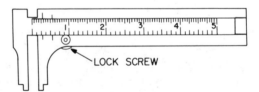

Figure 4-10 The slide caliper.

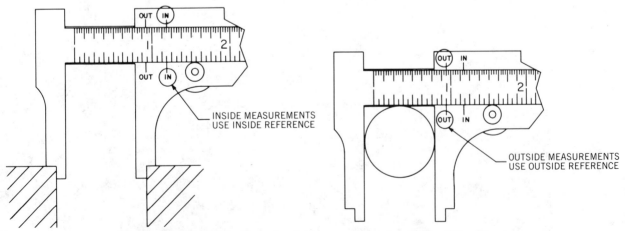

Figure 4-11 Making inside and outside measurements with the slide caliper.

ATTACHMENTS FOR STEEL RULES

Several attachments have been developed to extend the use of the steel rule. The most common attachments are keyseat clamps, rule clamps, right angle rule clamps, and drill point gages.

KEYSEAT CLAMPS

Keyseat clamps (Figure 4-12) are used to hold a steel rule in the proper position for marking a shaft for a keyseat. The proper use of this attachment is shown in Figure 4-13.

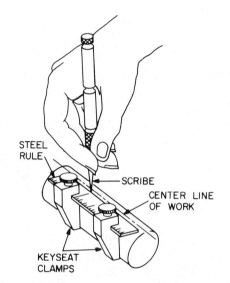

Figure 4-12 Keyseat clamps.

Figure 4-13 Proper use of the keyseat clamps.

RULE CLAMPS

Rule clamps (Figure 4-14) are used to hold two rules together when making a measurement beyond the range of either rule. These clamps are well suited for extending the range of rules commonly found in a machinist's toolbox. For example, if you had to measure a part that was 16 in. long and the only rules you had in your toolbox were a 12-in. rule and a 6-in. rule, all you would have to do is clamp your rules together with a rule clamp and make the measurement. These clamps are capable of holding most rules commonly found in a machine shop, but they can accept only rule widths between $\frac{1}{2}$ and $1\frac{1}{4}$ in.

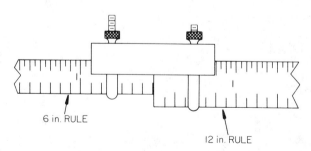

Figure 4-14 The rule clamp.

RIGHT ANGLE RULE CLAMPS

The right angle rule clamp (Figure 4-15) is used to hold two rules at exactly 90° to each other. Typical uses include measuring depths of large slots (Figure 4-16).

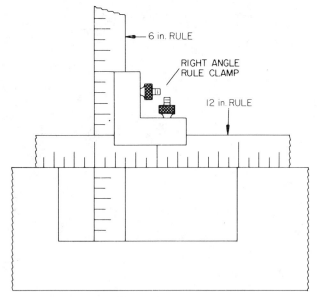

Figure 4-16 Using a right angle rule clamp.

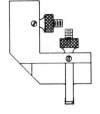

Figure 4-15 The right angle rule clamp.

DRILL POINT GAGES

The drill point gage (Figure 4-17) is used to help a machinist sharpen drills. The gage is simply placed on a steel rule and clamped in place. The 59° angle is used to check all general-purpose drill points. The graduated lines along the edge are used to help make sure the drill point is located in the center of the drill. Figure 4-18 shows how this attachment is used.

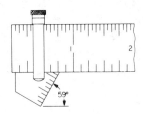

Figure 4-17 Drill point gage.

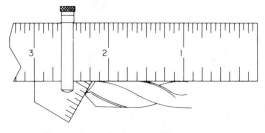

Figure 4-18 Using a drill point gage.

SELF TEST

1. List the three primary types of adjustable squares.
2. Which adjustable square is the most versatile and can do the most?
3. Which adjustable square can be used to check slight angles?
4. Which adjustable square has a center head?

5. Which adjustable square can be used to check the angle of a drill point?
6. Which type of steel rule is used with a depth gage?
7. What part features are commonly measured with a depth gage?
8. What type of measurements can be made with a slide caliper?
9. List four attachments that can extend the versatility of a steel rule.
10. Identify parts a to g in Figure 4-19.

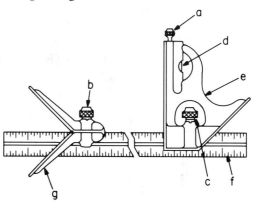

Figure 4-19

Answers to Self Test

1. a. Combination square
   b. Double square
   c. Diemakers' square
2. Combination square
3. Diemakers' square
4. Combination square
5. Double square
6. Narrow rule
7. Depth of slots, holes, and shoulders
8. Inside and outside
9. a. Keyseat clamps
   b. Rule clamps
   c. Right angle rule clamps
   d. Drill point gage
10. a. Part a is the scribe
    b. Part b is the locking nut
    c. Part c is the locking nut
    d. Part d is the spirit level
    e. Part e is the square head
    f. Part f is the steel rule
    g. Part g is the center head

SECTION III

PRECISION MEASURING TOOLS

UNIT 5 **Micrometers**

UNIT 6 **Verniers**

UNIT 5 # Micrometers

The micrometer is the workhorse of the measuring tools in the machine shop. Micrometers are used to measure external features, internal features, and depths. In addition, several variations of the standard micrometer make the usefulness of this tool almost limitless. The micrometer can make direct measurements to 0.0001 in., or 0.002 mm. To make accurate measurements, you must know how to use and care for this tool.

CONSTRUCTION OF A MICROMETER

To make the best use of a micrometer, it helps to know its parts. The micrometer, shown in Figure 5-1, consists of a *frame* that holds the *anvil* and *spindle*. The anvil and spindle make contact with the workpiece during a measurement. On the other side of the frame is the *sleeve*. The sleeve contains the graduations dividing an inch into 40 equal parts. On the outside of the sleeve is the *thimble*. The thimble further divides the inch into 0.001 in. increments. At the end of the thimble is the *ratchet stop*. The ratchet stop is used to tighten the spindle against the workpiece. As soon as enough pressure is applied, the ratchet will slip. When the ratchet slips, tighten the lock nut, remove the micrometer, and take your reading.

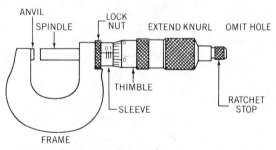

Figure 5-1 Parts of a micrometer.

HOW A MICROMETER WORKS

The micrometer works on the principle of the screw thread. In technical terms, all linear movement of the micrometer spindle is controlled by the rotational movement of a screw thread. Figure 5-2 shows how this rotary movement of the thimble is converted into linear travel of the spindle.

The rate at which the bolt advances is controlled by the number of threads per inch. If a bolt has 10 threads per inch, for instance, it would take 10

43

44 SECTION III PRECISION MEASURING TOOLS

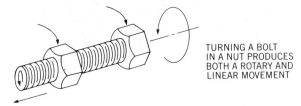

Figure 5-2 Screw thread principle of transfering rotary motion to linear travel.

revolutions to move the bolt 1 in. If the bolt were only rotated one revolution, it would move only $\frac{1}{10}$ in. If we convert this to a decimal, it becomes 0.100 in. and represents the *pitch* of the thread. The pitch of a thread is distance from a point on one thread to the same point on the next thread (Figure 5-3). From this you can see that one revolution of a bolt with 10 threads per inch results in 0.100 in. of linear travel. If a bolt had 20 threads per inch its movement per revolution would equal $\frac{1}{20}$ in., or 0.050 in.

Figure 5-3 The pitch of a thread.

The micrometer has 40 threads per inch. This means the thread will advance the spindle 0.025 in. with each revolution of the thimble. With 40 revolutions the spindle will advance a full inch. The reading line on the sleeve is divided into increments of 0.025 in. Each time the thimble is rotated a full revolution, the spindle moves from one line to the next (Figure 5-4). These lines are numbered at every 0.100 in. increment. The line marked 1 is equal to 0.100 in, the line marked 2 is equal to 0.200 in., and so on. The smaller lines between the numbered lines are equal to 0.025 in. Reading left to right, these lines are equal to 0.025, 0.050, 0.075, 0.125, 0.150, 0.175 in., and so on (Figure 5-5).

The graduations on the front edge of the thimble are used to subdivide the sleeve readings. These graduations divide the thimble into 25 equal parts, each equal to 0.001 in. When reading a micrometer, it is a simple matter to add these graduations to get your measurement.

Figure 5-4 Thimble movement.

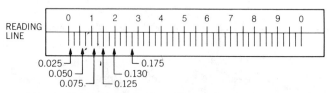

Figure 5-5 Sleeve graduations.

READING A MICROMETER TO 0.001 IN.

The first step in reading a micrometer is learning to use the ratchet stop or friction thimble (Figure 5-6). Both the ratchet stop and friction thimble prevent overtightening the spindle. Rely on these two devices. They can help you get consistently accurate readings. Some micrometers have neither a friction thimble or ratchet stop. With these you will have to develop a feel for the correct reading. The easiest way to do this is to check the micrometer against its standard and note the feel of the micrometer against the standard. Now match this feel against the part and you will have an accurate reading.

RATCHET STOP   FRICTION THIMBLE

Figure 5-6 Rachet stop and friction thimble are used to insure correct measuring pressure.

To read a micrometer in thousandths (0.001), first count the number of exposed graduation lines on the sleeve and multiply by 0.025 in. Next add the number on the thimble to the number on the sleeve, and your reading is complete (Figure 5-7).

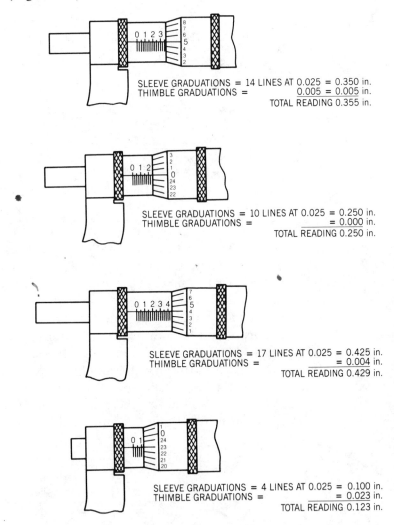

SLEEVE GRADUATIONS = 14 LINES AT 0.025 = 0.350 in.
THIMBLE GRADUATIONS =   0.005 = 0.005 in.
TOTAL READING 0.355 in.

SLEEVE GRADUATIONS = 10 LINES AT 0.025 = 0.250 in.
THIMBLE GRADUATIONS =   = 0.000 in.
TOTAL READING 0.250 in.

SLEEVE GRADUATIONS = 17 LINES AT 0.025 = 0.425 in.
THIMBLE GRADUATIONS =   = 0.004 in.
TOTAL READING 0.429 in.

SLEEVE GRADUATIONS = 4 LINES AT 0.025 = 0.100 in.
THIMBLE GRADUATIONS =   = 0.023 in.
TOTAL READING 0.123 in.

Figure 5-7 Reading a micrometer to 0.001 in.

46 SECTION III PRECISION MEASURING TOOLS

Now that you know how to read a micrometer to 0.001 in., let's try some practice problems (see Figure 5-8).

Sample Problem 1

Read the 15-micrometer settings and place the correct answer beneath the diagram.

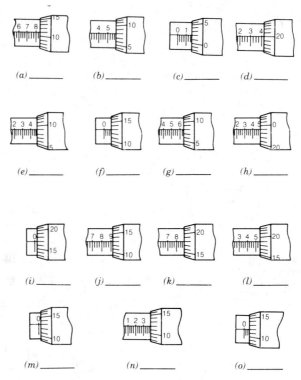

Figure 5-8 Practice problem 1.

READING A MICROMETER TO 0.0001 IN.

Some micrometers can be read to 0.0001 in. These micrometers are similar to 0.001 in. micrometers, but they have a vernier scale on the top of the sleeve (Figure 5-9). This vernier scale is used to divide each 0.001 in. graduation into 10 equal parts, or 0.0001 in.

When reading a 0.0001 in. micrometer, first read the micrometer to the thousandths as you have earlier. Next, turn the micrometer so the vernier scale is clearly visible. Notice the 11 lines in the vernier scale—0, 1, 2, 3, 4, 5, 6, 7, 8, 9, and 0 (Figure 5-10). Since there are two zero lines, we can eliminate one for the purpose of this explanation. Starting with the lower zero and counting upward to 9, you will notice there are 10 lines. On the thimble, there are only 9 lines. This feature, 10 lines occupying the same space as 9 lines, is the prin-

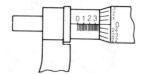

Figure 5-9 The vernier scale.

Unit 5   Micrometers   47

Figure 5-10 The vernier principle.

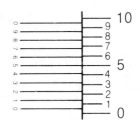

Figure 5-11 Only one set of lines will line up at any one time.

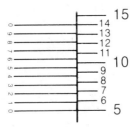

Figure 5-12 Only the zeros will line up together.

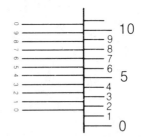

Figure 5-13 Number 6 shown in alignment with thimble graduation.

ciple behind the vernier scale system. Only one set of lines will line up the two 5 lines at any one time (Figure 5-11). The exception to this is the zero. If one zero is aligned, the other zero will also be aligned to a line on the thimble (Figure 5-12). This is why we dropped one zero in this discussion.

To read this scale, first find the two lines that are either aligned or the closest to being aligned (Figure 5-13). You will notice that the number 6 is the vernier scale line that aligns with a thimble line. This means that 0.0006 in. should be added to the thousandths reading. One point to remember when reading a vernier scale is: always use the number value shown on the vernier scale, not that shown on the thimble.

Now let's put it all together and make a complete reading to an accuracy of 0.0001 in. (Figure 5-14).

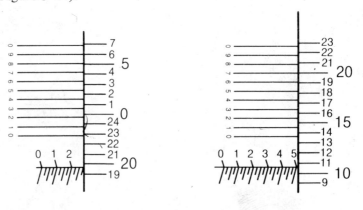

SLEEVE GRADUATIONS   =   10 LINES AT 0.025 =   0.250   in.
THIMBLE GRADUATIONS =   0.019   in.
VERNIER GRADUATION =   0.0009 in.
                        TOTAL READING 0.2699 in.

SLEEVE GRADUATIONS   =   20 LINES AT 0.025 =   0.500   in.
THIMBLE GRADUATIONS =   0.010   in.
VERNIER GRADUATION =   0.0007 in.
                        TOTAL READING 0.5107 in.

Figure 5-14 Reading a micrometer to 0.0001 in.

SECTION III PRECISION MEASURING TOOLS

Now that you know how to read a micrometer to 0.0001 in., let's try a few practice problems (see Figure 5-15).

Practice Problem 2

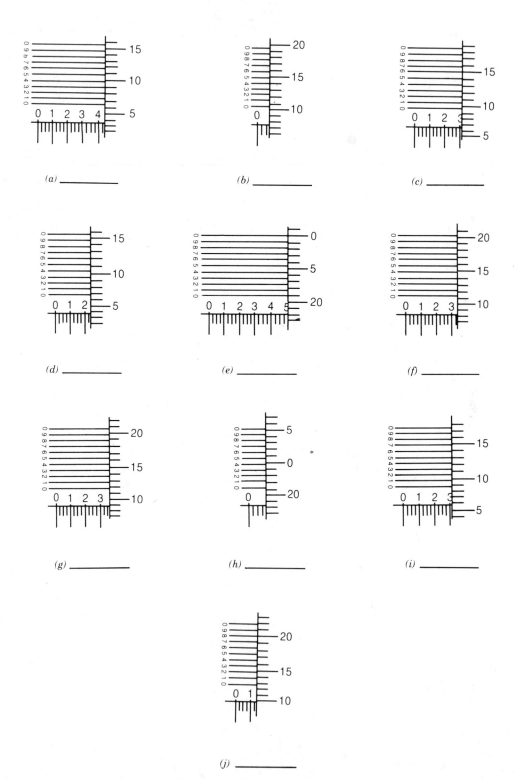

Figure 5-15 Practice problem 2.

READING A MICROMETER TO 0.01 mm

If you can read a micrometer to 0.001 in., you can read one to 0.01 mm. The main difference is in the use of millimeters instead of inches. The millimeter micrometer (Figure 5-16) has a thread with a pitch of 0.5 mm, so each revolution of the thimble will move the spindle $\frac{1}{2}$ mm. The sleeve on a millimeter micrometer is graduated in $\frac{1}{2}$ mm increments. The thimble is divided into 50 equal parts, so one line on the thimble is equal to $\frac{1}{50}$ revolution, or 0.5 mm/50 = 0.01 mm.

To read a millimeter micrometer, simply use the same process as used for the inch micrometer. First, count all the exposed lines on the sleeve and multiply by 0.5 mm. Next, add the value shown on the thimble to the sleeve reading (Figure 5-17).

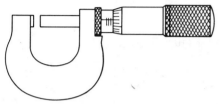

Figure 5-16 Millimeter micrometer.

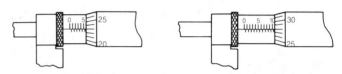

Figure 5-17 Reading a micrometer to 0.01 mm.

Now let's try a few practice problems (see Figure 5-18).

Practice Problem 3

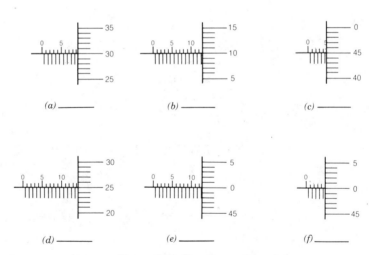

Figure 5-18 Practice problem 3.

READING A MICROMETER TO 0.002 mm

Reading micrometers that read directly to 0.002 mm requires just a slight adjustment from reading a 0.0001-in. micrometer. The vernier scale on this micrometer has graduated lines that are equal to 0.002 mm. When making a reading, simply follow the steps used for a 0.01-mm micrometer and add the vernier reading, as shown in Figure 5-19.

SECTION III PRECISION MEASURING TOOLS

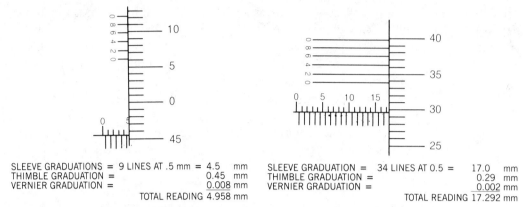

SLEEVE GRADUATIONS = 9 LINES AT .5 mm = 4.5 mm
THIMBLE GRADUATION = 0.45 mm
VERNIER GRADUATION = 0.008 mm
TOTAL READING 4.958 mm

SLEEVE GRADUATION = 34 LINES AT 0.5 = 17.0 mm
THIMBLE GRADUATION = 0.29 mm
VERNIER GRADUATION = 0.002 mm
TOTAL READING 17.292 mm

Figure 5-19 Reading a micrometer to 0.002 mm.

Now let's try some practice problems (see Figure 5-20).

Practice Problem 4

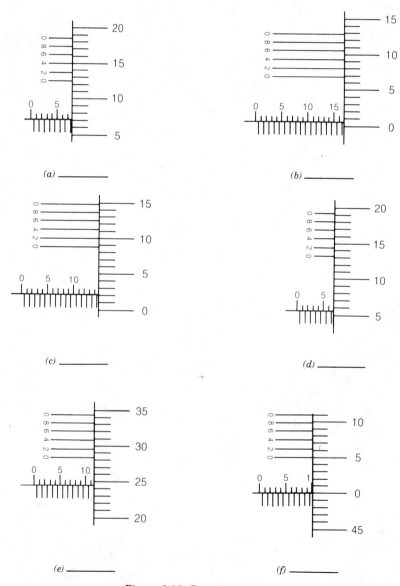

Figure 5-20 Practice problem 4.

TYPES OF MICROMETERS

There are three basic types of micrometers: outside, inside, and depth. Beyond these basic styles are several variations that greatly extend the use of the micrometer. Regardless of the style, the basic principles for reading a micrometer are the same.

OUTSIDE MICROMETERS

The standard outside micrometer is the most common type of micrometer found in the machine shop, and it is best suited for making all general-purpose measurements. The standard outside micrometer has a cylindrical anvil and spindle that are generally 0.250 in. in diameter (Figure 5-21). The most common variations of the outside micrometer have anvil and spindle combinations made for specific applications (Figure 5-22). These include the disc micrometer, blade micrometer, spline micrometer, screw thread micrometer, ball anvil micrometer, point micrometer, tube micrometer, and V-anvil micrometer.

The standard outside micrometer is available in ranges of 0 to 12 in. (0 to 300 mm) in 1-in. (25-mm) increments. Larger micrometers are available, but it is not likely you will ever use one larger than 12 in. (300 mm) in the normal machine shop. The specialty micrometers are generally made in only the 0 to 1 in. (0 to 25 mm) and 1 to 2 in. (25 to 50 mm) sizes.

Other variations of the standard outside micrometer are shown in Figure 5-23. These include the interchangeable anvil micrometer, deep throat micrometer, hub micrometer, indicating micrometer, and digital micrometer.

Figure 5-21 Standard spindle and anvil.

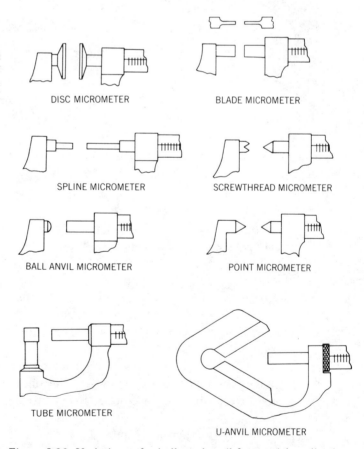

Figure 5-22 Variations of spindle and anvil for special applications.

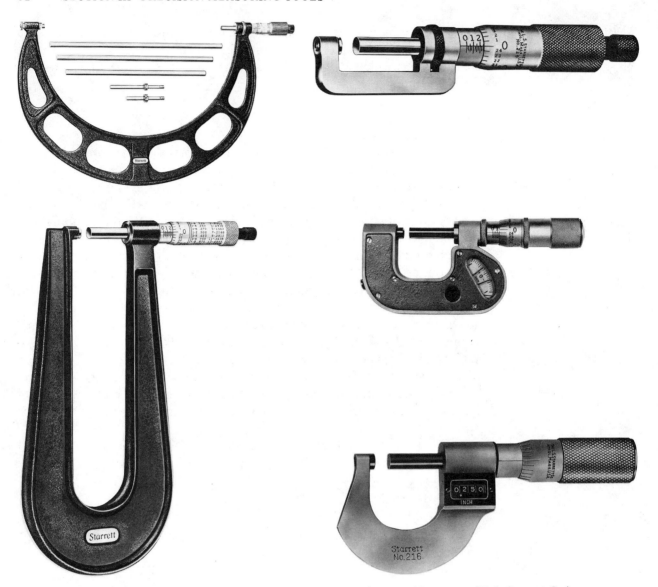

Figure 5-23 Variations of the standard outside micrometer. (Courtesy of L.S. Starrett Co.)

INSIDE MICROMETERS

The inside micrometer is designed to make internal measurements. The basic inside micrometer consists of a micrometer head and an assortment of different length rods that attach to the micrometer head (Figure 5-24). The normal range of an inside micrometer is $1\frac{1}{2}$ to 12 in. (40 to 300 mm). Again, larger sets are available, but they are seldom used in most machine shops. The micrometer head on most inside micrometers has a travel of only $\frac{1}{2}$ in. Variations of the standard inside micrometer include jaw-type inside micrometers and internal micrometers (Figure 5-25).

Reading the jaw-type inside micrometer is different from reading a standard inside micrometer. Jaw-type inside micrometers are read the reverse of standard micrometers. Instead of reading the sleeve graduations that are exposed, with a jaw-type inside micrometer you must read the sleeve graduations that are covered by the thimble (Figure 5-26).

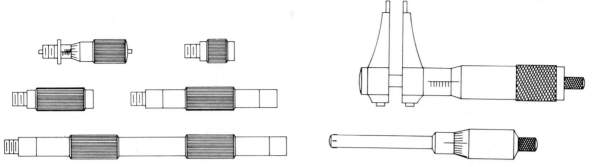

Figure 5-24 Inside micrometer.

Figure 5-25 Variations of the inside micrometer.

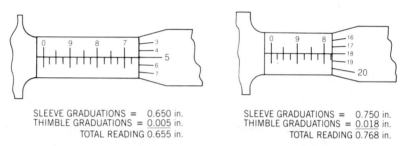

SLEEVE GRADUATIONS = 0.650 in.
THIMBLE GRADUATIONS = 0.005 in.
TOTAL READING 0.655 in.

SLEEVE GRADUATIONS = 0.750 in.
THIMBLE GRADUATIONS = 0.018 in.
TOTAL READING 0.768 in.

Figure 5-26 Reading the jaw-type inside micrometer.

DEPTH MICROMETERS

The depth micrometer (Figure 5-27) is designed to measure the depth of holes, slots, shoulders, and steps. This micrometer has a precision lapped base that is held against the reference surface while the micrometer is brought into contact with the workpiece at the measured point.

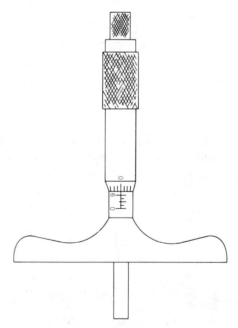

Figure 5-27 Depth micrometer.

54 SECTION III PRECISION MEASURING TOOLS

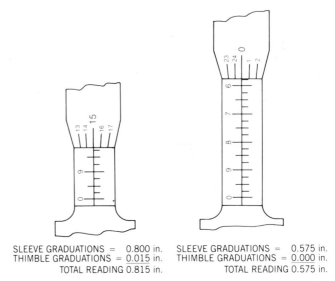

SLEEVE GRADUATIONS = 0.800 in. SLEEVE GRADUATIONS = 0.575 in.
THIMBLE GRADUATIONS = 0.015 in. THIMBLE GRADUATIONS = 0.000 in.
TOTAL READING 0.815 in. TOTAL READING 0.575 in.

Figure 5-28 Reading the depth micrometer.

When reading a depth micrometer, make your reading in the same way as for a jaw-type inside micrometer. You must read all the sleeve graduations that are covered by the thimble. This process is shown in Figure 5-28.

Depth micrometers are available in ranges of 0 to 12 in. (0 to 300 mm) in 1-in. (25-mm) increments. Each 1-in. (25-mm) increment of measurement requires a different rod, so with a 0 to 12 in. (0 to 300 mm) depth micrometer there are 12 rods. Rods are changed in a depth micrometer, as shown in Figure 5-29.

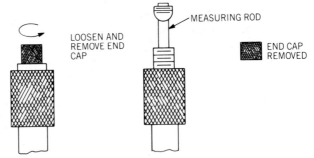

Figure 5-29 Changing measuring rods with a depth micrometer.

MAKING MEASUREMENTS WITH A MICROMETER

Measurements made with a micrometer, like a steel rule, can be divided into two categories, direct and indirect. Direct measurements are made with the micrometer alone. Indirect, or transfer, measurements are made with a transfer measuring tool along with a micrometer.

DIRECT MEASUREMENTS

When making direct measurements with a micrometer, remember these points.

1. Make sure the workpiece is clean and free of chips and dirt.

2. Make sure the spindle and anvil are clean. The proper size rod must be cleaned before mounting it in the micrometer.

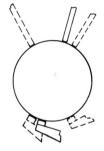

Figure 5-30 Rocking micrometer to insure good contact.

3. When tightening the micrometer against the workpiece, use the friction thimble or ratchet stop.
4. Make sure the anvil and spindle are squarely seated. If necessary, move the micrometer to insure a good contact (Figure 5-30).
5. Once the measurement is made, tighten the lock nut.
6. To be sure the thimble does not move, remove the micrometer carefully.
7. Make the reading.
8. Recheck the measurement. If you get the same reading, you can trust your measurement. If you get a different reading, you must go back, check each step of the process, and make another measurement.

The most important thing is to be sure you have proper contact between the workpiece and the contact points of the measuring tool. Incorrect contact will always result in incorrect measurements. A few areas where errors can occur are shown in Figure 5-31.

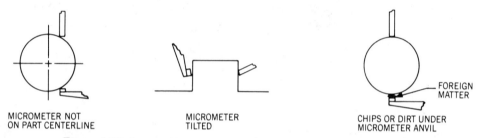

Figure 5-31 Inaccurate contacts result in inaccurate measurements.

TRANSFER MEASUREMENTS

The transfer measuring tools most often used with a micrometer are the small hole gage and the telescoping gage. These tools are adjusted to the size of the feature being measured and then locked. Once removed, they can be accurately measured with a micrometer.

Small Hole Gages. Small hole gages, shown in Figure 5-32, are available in sets of four gages ranging from 0.125 to 0.500 in. (3.2 to 12.7 mm). The measuring end is first inserted into the hole and the knurled end is tightened. The gage is removed when the correct amount of drag between the workpiece and gage is obtained. The outside micrometer is then used to measure the gage. Again, always duplicate the drag between the hole and the gage. When the drag between the micrometer and the gage feels the same as that in the hole, you can make your reading.

Telescoping Gages. Telescoping gages, shown in Figure 5-33, are available in sets of six gages ranging from 0.312 to 6.000 in. (7.9 to 152.4 mm). These gages operate with spring loaded contact points that expand against the inside of a hole. The position of the contacts is locked by turning the knurled nut on the end of the gage handle.

With a telescoping gage, first depress the contacts with your fingers. When the distance across the contacts is slightly smaller than the opening you are measur-

56 SECTION III PRECISION MEASURING TOOLS

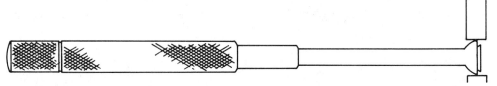

Figure 5-32 Small hole gages. (Courtesy of Precision Brand Products, Inc.)

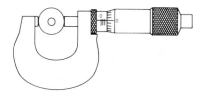

Figure 5-34 Checking the micrometer with a standard.

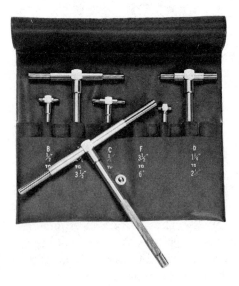

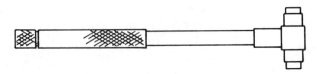

Figure 5-33 Telescoping gages. (Courtesy of Precision Brand Products, Inc.)

ing, lock the gage. Now insert the gage and loosen the knurled knob. Never collapse the gage completely and snap the contacts into the workpiece since this could cause damage to the gage and result in an inaccurate measurement. When the contacts are against the workpiece, slightly tighten the lock. Try to move the gage left and right and back and forth. If no movement occurs, the gage is properly positioned. If the gage moves, simply loosen the lock and allow the contacts to expand. When you have no movement, tighten the lock and carefully remove the gage. You should have a slight amount of drag between the gage and the workpiece. Once the gage is removed, make your measurement with a micrometer using the same process as that for the small hole gage.

Notice that a great deal has been said about the drag between the workpiece and the measuring tool. Again, this drag is referred to as "feel" and is essential in making precision measurements. As you become experienced in making measurements, you will develop this sense of feel. To help you learn the correct pressure, make several measurements using the ratchet stop or friction thimble and note the amount of pressure between the micrometer and the workpiece. Now try to duplicate this pressure without the friction thimble or ratchet stop. When you can match the pressure, you are on your way to developing feel.

CARE OF MICROMETERS

The following are a few tips about using your micrometer.

1. Treat a micrometer carefully. Do not spin the thimble to open or close the micrometer. Turn the thimble gently.

2. Do not leave a micrometer on a machine table or workbench. Put it in your apron pocket or toolbox or in its case. When stored, be sure to coat the instrument with a thin film of oil or corrosion-resistant compound.

3. Never measure a moving part or measure close to moving cutters.

4. Periodically check the accuracy of your micrometer against a standard (Figure 5-34) to be sure the micrometer is measuring accurately.

SELF TEST

1. Identify the parts of the micrometer labeled A to G in Figure 5-35.

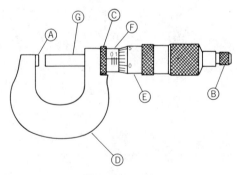

Figure 5-35

2. What is the value of the sleeve graduations on an inch micrometer? (Circle one.)

   a. 0.100 in.

   b. 0.050 in.

   c. 0.001 in.

   d. 0.025 in.

3. What is the value of the thimble graduations on an inch micrometer? (Circle one.)

   a. 0.001 in.

   b. 0.010 in.

   c. 0.0001 in.

   d. 0.100 in.

4. What is the value of the vernier scale graduations on an inch micrometer? (Circle one.)

   a. 0.002 in.

   b. 0.0001 in.

   c. 0.001 in.

   d. 0.010 in.

5. What is the value of the thimble graduations on a millimeter micrometer? (Circle one.)

   a. 0.0001 mm

   b. 0.025 mm

   c. 0.01 mm

   d. 0.001 mm

6. What is the value of the vernier scale graduations on a millimeter micrometer? (Circle one.)

   a. 0.0001 mm

   b. 0.001 mm

   c. 0.002 mm

   d. 0.020 mm

7. What is the value of the sleeve graduations on a millimeter micrometer? (Circle one.)

   a. 0.05 mm

   b. 0.005 mm

   c. 5.0 mm

   d. 0.5 mm

8. Which of the following is *not* one of the basic micrometers? (Circle one.)

   a. Outside micrometer

   b. Standard micrometer

   c. Depth micrometer

   d. Inside micrometer

9. Which of the following is an example of a variation of the outside micrometer? (Circle one.)

   a. Sleeve micrometer

   b. Thread micrometer

   c. Jaw micrometer

   d. Standard micrometer

10. Which of the following is a transfer measuring tool generally used with a micrometer? (Circle one.)

    a. Small diameter gage

    b. Telephonic gage

    c. Telescoping gage

    d. Small bore gage

11. What is the term used to describe the drag between the workpiece and the measuring tool? (Circle one.)

    a. Drag

    b. Friction

    c. Pressure

    d. Feel

12. Read the micrometers in Figure 5-36 and write the correct answer in the spaces provided.

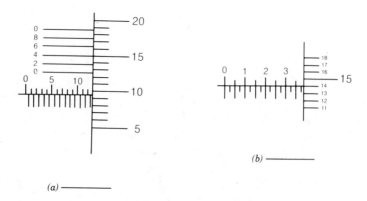

Figure 5-36

SECTION III PRECISION MEASURING TOOLS

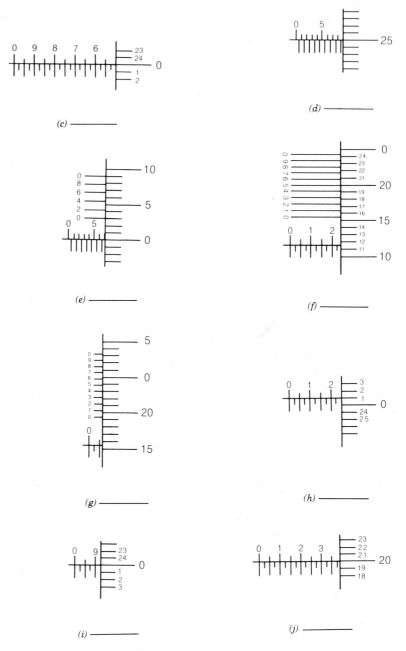

Figure 5-36 (continued)

ANSWERS TO SELF TEST

1. a. Part A is the anvil
   b. Part B is the ratchet stop
   c. Part C is the lock nut
   d. Part D is the frame
   e. Part E is the thimble
   f. Part F is the sleeve
   g. Part G is the spindle
2. d
3. a
4. b
5. c
6. c
7. d
8. b
9. b
10. c

11. d
12. a. 12.594 mm
    b. 0.389 in.
    c. 0.500 in.
    d. 8.75 mm
    e. 7.000 mm
    f. 0.2365 in.
    g. 0.0654 in.
    h. 0.251 in.
    i. 0.875 in.
    j. 0.395 in.

Answers to Practice Problems

1. a. 0.836 in.
   b. 0.583 in.
   c. 0.152 in.
   d. 0.420 in.
   e. 0.484 in.
   f. 0.088 in.
   g. 0.633 in.
   h. 0.499 in.
   i. 0.017 in.
   j. 0.913 in.
   k. 0.893 in.
   l. 0.544 in.
   m. 0.013 in.
   n. 0.387 in.
   o. 0.062 in.

2. a. 0.4283 in.
   b. 0.0574 in.
   c. 0.3066 in.
   d. 0.2286 in.
   e. 0.4930 in.
   f. 0.3333 in.
   g. 0.3587 in.
   h. 0.0932 in.
   i. 0.3055 in.
   j. 0.1348 in.

3. a. 9.30 mm
   b. 12.60 mm
   c. 4.95 mm
   d. 14.25 mm
   e. 12.50 mm
   f. 5.00 mm

4. a. 7.574 mm
   b. 16.502 mm
   c. 14.522 mm
   d. 6.558 mm
   e. 11.246 mm
   f. 10.000 mm

UNIT 6 # Verniers

The vernier, like the micrometer, is a versatile and useful measuring tool. Verniers are sliding scale measuring instruments. They are accurate to 0.001 in., or 0.02 mm. There are several variations of the basic vernier principle, but we will limit our discussion to the tools commonly found in the machine shop.

CONSTRUCTION OF A VERNIER

To understand how to use a vernier instrument, it helps to learn the parts of a vernier caliper. Although the vernier caliper is only one of a variety of vernier instruments, it shows how most vernier instruments are constructed.

As you can see in Figure 6-1, the vernier caliper consists of a beam that holds the *solid jaw* and the *movable jaw*. Contained in the movable jaw are the *lock screws* and the *vernier plate*. On some models of verniers, there is also a fine adjustment feature that consists of a *fine adjustment slide* and a *fine adjustment nut*. At the ends of the solid and movable jaws are the *contact points*. The measurement contact points are made in two basic styles (Figure 6-2). One contains both the inside and outside contacts at the end of each jaw. The other uses two separate sets of jaws. Regardless of the design, all vernier-type instruments work in basically the same way.

The vernier works on a principle of sliding scales. The beam of the vernier contains the main scale, and the movable jaw contains the vernier scale. Each of these scales has a different scale discrimination; however, when read together, they can accurately measure to 0.001 in. (0.02 mm).

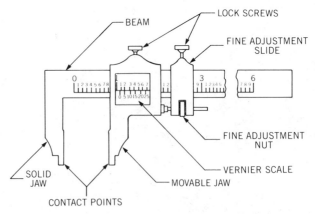

Figure 6-1 Parts of a vernier caliper.

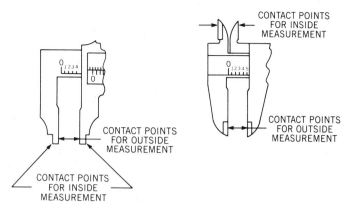

Figure 6-2 Measurement contact points.

As stated in Unit 5, when two sets of scales with different graduations occupy the same space, only one set of lines will line up at any one time (Figure 6-3). This same principle is true for all vernier instruments.

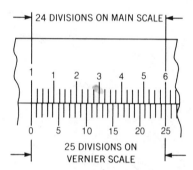

Figure 6-3 The vernier principle.

READING A 25-DIVISION VERNIER

When reading any vernier, first identify the scale on the instrument. In the case of a 25-division vernier, there are 25 divisions on the vernier plate and 40 divisions per inch on the main scale.

The main scale on a 25-division vernier instrument is graduated, as shown in Figure 6-4. Notice that these graduations are the same as those used on the sleeve of a micrometer. There are 40 graduations per inch on the main scale. This means each line is equal to $\frac{1}{40}$ in., or 0.025 in. Each small, numbered line is equal to 0.100 in., and each small, unnumbered line is equal to 0.025 in., again the same as a micrometer. The large numbered lines represent whole inch units.

When reading the main scale graduations, all lines must be considered in the reading. For example, the readings shown in Figure 6-5 must be read as 1.100 in., 1.375 in., and 2.125 in., not 0.100 in., 0.375 in., and 0.125 in. Whole inch units must always be included in any reading over 1 in.

The divisions on the vernier scale are used to subdivide further the main scale readings. Since the main scale graduations divide each inch into 40 parts and the vernier scale divides each part into 25 parts, each line on the vernier scale is

SECTION III PRECISION MEASURING TOOLS

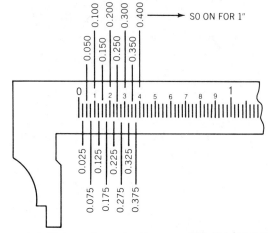

Figure 6-4 Main scale graduations on a 25-division vernier.

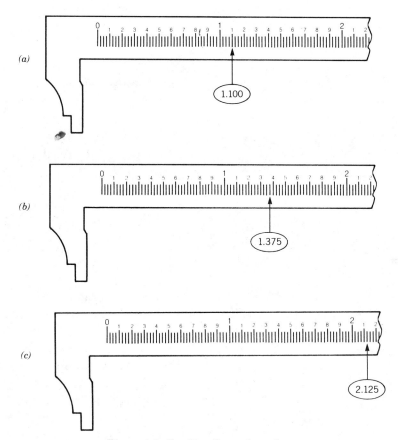

Figure 6-5 Reading the main scale.

equal to $\frac{1}{40} \times \frac{1}{25} = \frac{1}{1000}$, or 0.001 in. This is identical to the thimble graduations on a micrometer. Now let's put all of this together and make an actual reading with a vernier.

The first step in reading a 25-division vernier is to locate the position of the zero on the vernier plate with reference to the main scale. As shown in Figure 6-6, the zero is just to the right of the first line after the 2-in. graduation. This means the reading at this point is 2.025 in. Now find the one line on the vernier

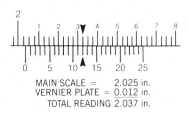

MAIN SCALE = 2.025 in.
VERNIER PLATE = 0.012 in.
TOTAL READING 2.037 in.

Figure 6-6 Locating the position of the zero.

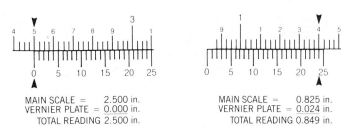

MAIN SCALE = 2.500 in.
VERNIER PLATE = 0.000 in.
TOTAL READING 2.500 in.

MAIN SCALE = 0.825 in.
VERNIER PLATE = 0.024 in.
TOTAL READING 0.849 in.

Figure 6-7 Reading the 25-division vernier.

plate that is lined up with a line on the main scale. In this case the number is 10 + 2, or 12. Adding this to the other reading, the total measurement is 2.037 in. Let's try a few more from Figure 6-7.

Now that you know how to read a 25-division vernier, here are a few practice problems (see Figure 6-8).

Practice Problem 1

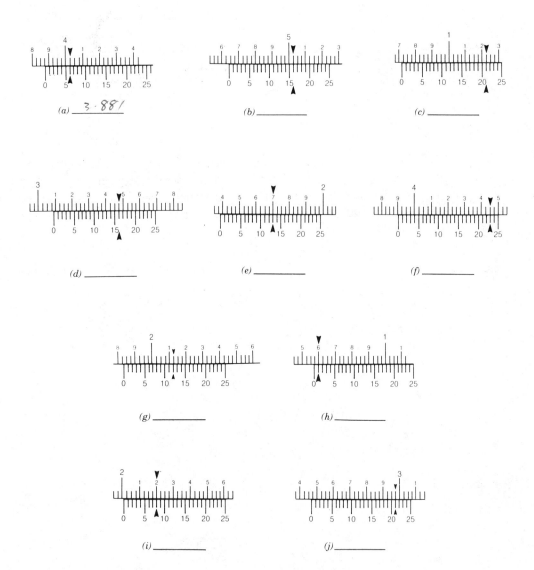

(a) 3.881

(b) _____

(c) _____

(d) _____

(e) _____

(f) _____

(g) _____

(h) _____

(i) _____

(j) _____

Figure 6-8 Practice problem 1.

READING A 50-DIVISION VERNIER

The 50-division vernier is a variation of the 25-division type. The main reason for using this type of vernier is its increased readability. The 25-division vernier is sometimes difficult to read because of the close spacing of the graduations. The 50-division vernier spreads the lines out over a larger area and makes reading this instrument much easier.

The main scale on the 50-division vernier has graduations spaced as shown in Figure 6-9. The main scale has 20 divisions per inch instead of the 40 divisions per inch of the 25-division type. The vernier scale has 50 divisions, each equal to $\frac{1}{20} \times \frac{1}{50} = \frac{1}{1000}$, or 0.001 in.

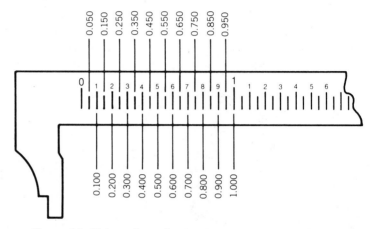

Figure 6-9 Main scale graduations on a 50-division vernier.

Reading a 50-division vernier is basically the same process as that for the 25-division type. First, locate the position of the zero with reference to the main scale. Next, find the vernier plate line, which lines up with a line on the main scale. Add the readings together and you have your measurement (Figure 6-10).

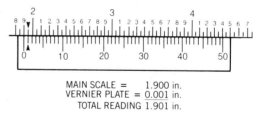

MAIN SCALE = 1.900 in.
VERNIER PLATE = 0.001 in.
TOTAL READING 1.901 in.

Figure 6-10 Reading a 50-division vernier.

Now that you can read a 50-division vernier, let's try a few more practice problems (see Figure 6-11).

Practice Problem 2

(a) _____

Figure 6-11 Practice problem 2.

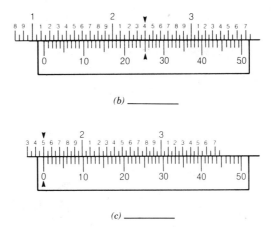

(b) _____

(c) _____

Figure 6-11 (*continued*)

READING A MILLIMETER VERNIER

Reading a millimeter vernier is basically the same process as reading an inch vernier. The differences are the graduations on the main scale and the vernier plate. The main scale on a millimeter vernier is graduated in whole millimeter units and the vernier plate is divided into 50 parts, each equal to $\frac{1}{50}$ mm, or 0.02 mm.

When reading a millimeter vernier, the first step is to locate the position of the zero on the vernier scale with reference to the graduations on the main scale. Next, find the one line on the vernier scale that is in alignment with a line on the main scale. Add the two values together. As shown in Figure 6-12, the zero is slightly past the 102-mm line. Now, looking down the vernier scale, you should see the 0.66-mm line aligned with a line on the main scale. Together this reading is 102.66 mm. Again, let's try a few practice problems (see Figure 6-13).

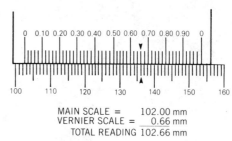

MAIN SCALE = 102.00 mm
VERNIER SCALE = 0.66 mm
TOTAL READING 102.66 mm

Figure 6-12 Reading a millimeter vernier.

Practice Problem 3

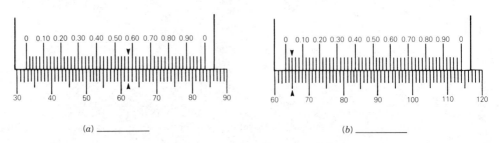

(a) _____   (b) _____

Figure 6-13 Practice problem 3.

68 SECTION III PRECISION MEASURING TOOLS

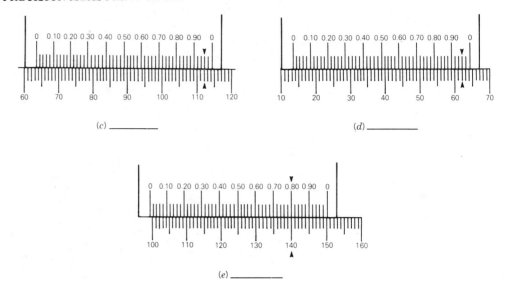

(c) _____

(d) _____

(e) _____

Figure 6-13 (*continued*)

READING DIAL-TYPE MEASURING TOOLS

Dial-type measuring tools are a cross between dial gages and vernier instruments. The most common type of dial measuring tool is the dial caliper (Figure 6-14). The dial caliper is quick, accurate, and can eliminate many of the reading errors that can occur with verniers.

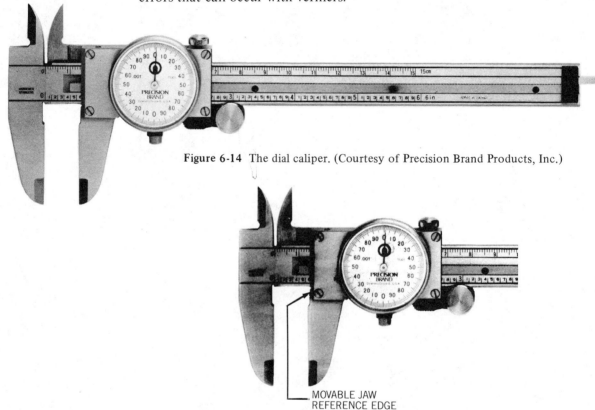

Figure 6-14 The dial caliper. (Courtesy of Precision Brand Products, Inc.)

Figure 6-15 Movable jaw reference edge. (Courtesy of Precision Brand Products, Inc.)

On any dial-type measuring tool, your first step is to locate the position of the movable jaw reference edge (Figure 6-15). This reference edge is used to measure the 0.100-in. increments marked on the main scale. The remaining reading is then taken directly from the dial. The reading shown in this figure is 0.601 in. The main scale graduations also indicate whole inch units. Always remember to include the whole inch increments in any reading.

The dials on most dial calipers are available in two styles (Figure 6-16). One dial has a range of 0.100 in., the other has a range of 0.200 in. Both dials are read the same way but, since the 0.200-in. dial has closer graduations, you will need to be more careful in reading the dial to be sure you have a correct measurement.

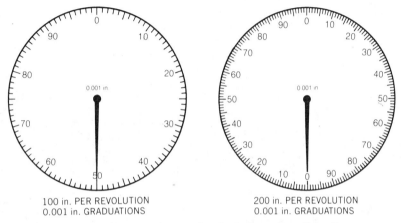

Figure 6-16 Dial variations for the inch reading dial caliper.

Dial calipers are also available with millimeter graduations. These calipers are accurate to either 0.02 or 0.05 mm and have dials similar to those shown in Figure 6-17. These calipers are read in the same way as the inch-type dial calipers.

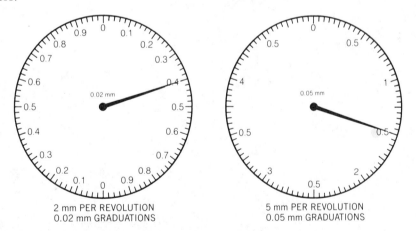

Figure 6-17 Millimeter reading dial calipers.

When reading any dial caliper, first locate the reference edge to find the first reading. This reading will either be in 0.100-in. units or 1-mm units. Next add this first reading to the value shown on the dial for the complete measurement (Figure 6-18). Now let's try a few practice problems (see Figure 6-19).

70 SECTION III PRECISION MEASURING TOOLS

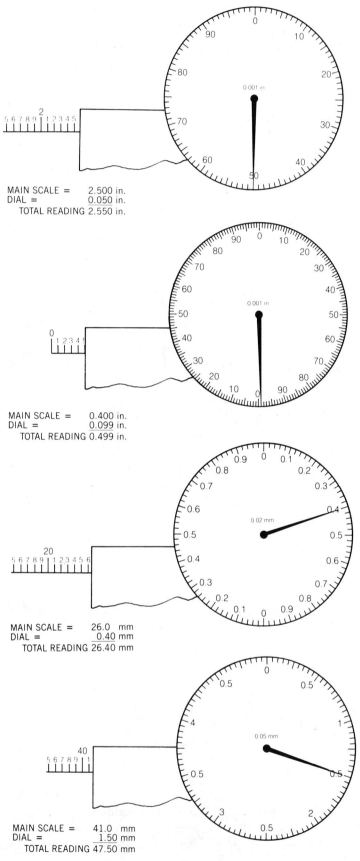

Figure 6-18 Reading the dial caliper.

Practice Problem 4

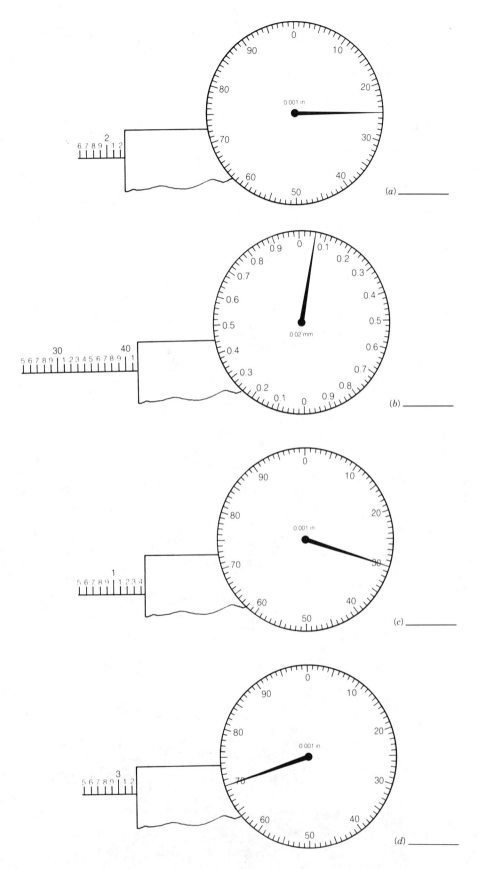

Figure 6-19 Practice problem 4.

72 SECTION III PRECISION MEASURING TOOLS

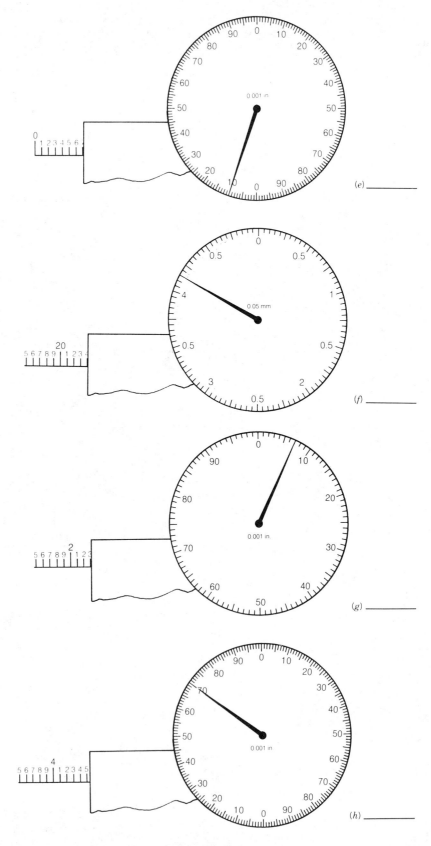

Figure 6-19 (continued)

TYPES OF VERNIERS

The vernier principle of measurement has been applied to a wide range of measuring tools. The tools you are most likely to use are the vernier caliper, vernier depth gage, vernier height gage, and vernier gear tooth caliper (Figure 6-20).

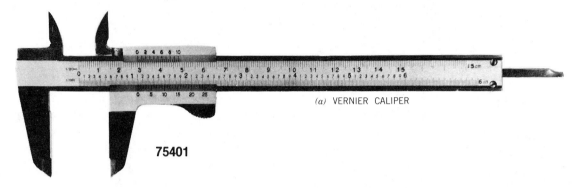

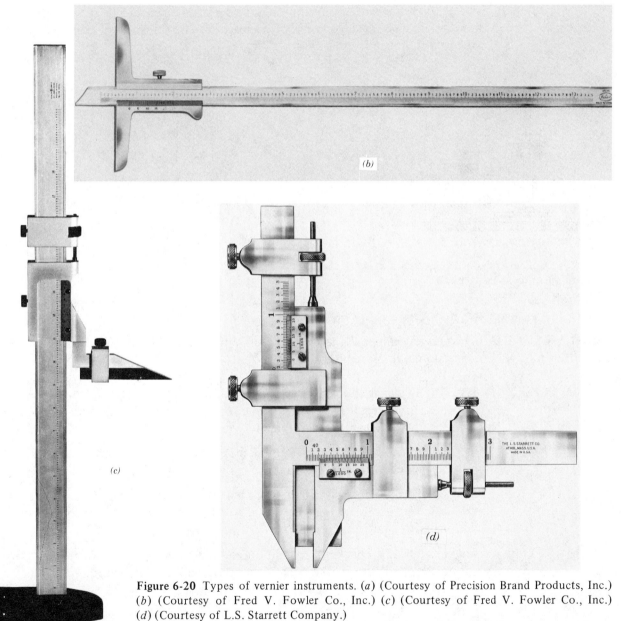

Figure 6-20 Types of vernier instruments. (*a*) (Courtesy of Precision Brand Products, Inc.) (*b*) (Courtesy of Fred V. Fowler Co., Inc.) (*c*) (Courtesy of Fred V. Fowler Co., Inc.) (*d*) (Courtesy of L.S. Starrett Company.)

74 SECTION III PRECISION MEASURING TOOLS

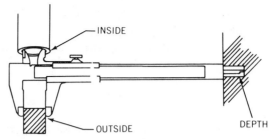

Figure 6-21 The vernier caliper used to measure inside, outside, and depth dimensions.

The vernier caliper is the most common type of vernier instrument used for inside and outside measurements. Models exist, however, that can also measure depths (Figure 6-21). The vernier depth gage is mainly used for measuring the depth of holes, slots, grooves, and steps.

Vernier height gages are used for precision layout and for checking sizes from a flat reference surface, such as a surface plate (Figure 6-22). The vernier gear tooth caliper is used exclusively for measuring the thickness of gear teeth along their pitch line (Figure 6-23).

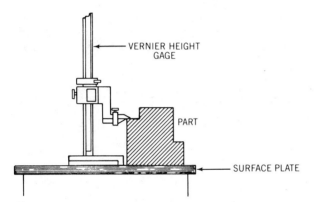

Figure 6-22 The vernier height gage used to check the size of a part on a surface plate.

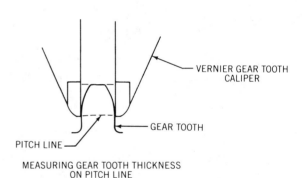

Figure 6-23 Measuring a gear tooth.

USING A VERNIER CALIPER The vernier caliper is the one type of vernier you will use most often. Here are a few points to remember about this tool.

1. The beam of the vernier caliper must always be held perpendicular to the measured surfaces (Figure 6-24). Any misalignment (Figure 6-25) will result in inaccurate measurements.

2. When making internal measurements, the beam must be held perpendicular to the measured surfaces and aligned with the part centerline (Figure 6-26).

3. Never apply excessive force when making a measurement. This could cause the vernier to spring out of shape and therefore result in an incorrect measurement (Figure 6-27).

4. Periodically check the condition and alignment of the vernier jaws. Any wear or abuse can easily damage a vernier and make accurate measurement impossible (Figure 6-28).

Unit 6   Verniers   75

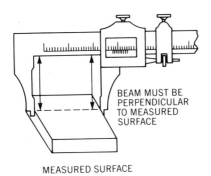

Figure 6-24 Holding the vernier caliper properly is important to the accuracy of any measurement.

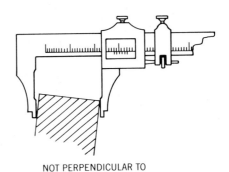

Figure 6-25 Misalignment results in incorrect readings.

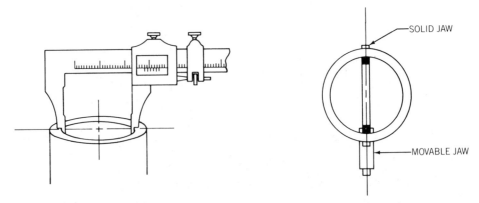

Figure 6-26 Using the vernier caliper for measuring internal part features.

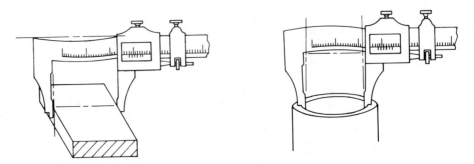

Figure 6-27 Never apply excessive force during a measurement.

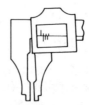

Figure 6-28 Worn or damaged jaws will affect the accuracy of any measurement.

76 SECTION III PRECISION MEASURING TOOLS

5. A vernier is a precision tool. Treat it carefully. Keep it properly stored in its case with a thin film of oil or corrosion-resistant compound.

SELF TEST

1. Identify parts A to I of the vernier caliper in Figure 6-29.

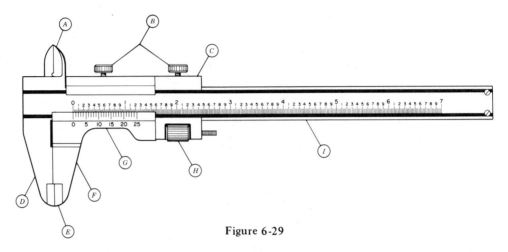

Figure 6-29

2. What type of measuring instrument is a vernier? (Circle one.)

   a. Screw thread

   b. Wedge lock

   c. Random scale

   d. Sliding scale

3. What is the value of the main scale graduations on a 25-division vernier? (Circle one.)

   a. 0.050 in.

   b. 0.250 in.

   c. 0.025 in.

   d. 0.0025 in.

4. What is the value of the main scale graduations on a 50-division vernier? (Circle one.)

   a. 0.001 in.

   b. 0.010 in.

   c. 0.025 in.

   d. 0.050 in.

5. What is the difference between a 25-division vernier plate and a 50-division vernier plate? (Circle one.)

   a. A 25-division vernier plate is more readable

   b. A 50-division vernier plate is more readable

c. A 25-division vernier plate is more accurate

d. A 50-division vernier plate is more accurate

6. How is the main scale graduated on a millimeter vernier? (Circle one.)

   a. 1 mm

   b. 0.01 mm

   c. 2 mm

   d. 0.02 mm

7. How closely can vernier instruments be read? (Circle one.)

   a. 0.002 in., or 0.001 mm

   b. 0.0001 in., or 0.02 mm

   c. 0.001 in., or 0.02 mm

   d. 0.010 in., or 0.05 mm

8. What is the first step in reading a dial caliper? (Circle one.)

   a. Locate the reference edge

   b. Zero the dial

   c. Locate the dial hand

   d. Align the dial hand and reference edge

9. What are the most common dial graduations used on inch-type dial calipers? (Circle one.)

   a. 50 and 100 divisions

   b. 100 and 200 divisions

   c. 50 and 75 divisions

   d. 25 and 50 divisions

10. Read the vernier instruments in Figure 6-30 and write the correct reading in the spaces provided.

(a)

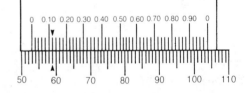

(b)

(c)

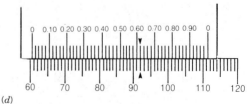

(d)

Figure 6-30

78 SECTION III PRECISION MEASURING TOOLS

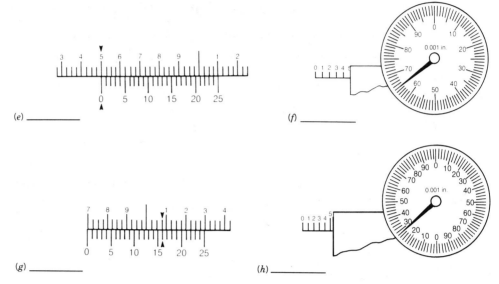

(e) _____ (f) _____

(g) _____ (h) _____

Figure 6-30 (*continued*)

Answers to Self Test

1. a. Part A is the inside measurement contact points
   b. Part B is the lock screws
   c. Part C is the fine adjustment slide
   d. Part D is the solid jaw
   e. Part E is the outside measurement contact points
   f. Part F is the movable jaw
   g. Part G is the vernier scale
   h. Part H is the fine adjustment nut
   i. Part I is the beam
2. d
3. c
4. d
5. b
6. a
7. c
8. a
9. b
10. a. 0.724 in.
    b. 5.12 mm
    c. 2.870 in.
    d. 6.62 mm
    e. 1.500 in.
    f. 0.465 in.
    g. 1.691 in.
    h. 1.528 in.

Answers to Practice Problems

1. a. 3.881 in.
   b. 4.641 in.
   c. 0.721 in.
   d. 3.091 in.
   e. 1.388 in.
   f. 3.898 in.
   g. 1.837 in.
   h. 0.576 in.
   i. 2.008 in.
   j. 2.471 in.

2. a. 3.105 in.
   b. 1.125 in.
   c. 1.500 in.
3. a. 32.58 mm
   b. 63.04 mm
   c. 63.96 mm
   d. 13.96 mm
   e. 99.80 mm

4. a. 2.225 in.
   b. 41.06 mm
   c. 1.430 in.
   d. 3.270 in.
   e. 0.710 in.
   f. 24.15 mm
   g. 2.307 in.
   h. 4.570 in.

SECTION IV
INDICATORS AND GAGES

UNIT 7	**Dial Indicators**
UNIT 8	**Gage Blocks and Fixed Gages**
UNIT 9	**Comparison Gages**
UNIT 10	**Levels, Straightedges, and Precision Squares**

UNIT 7

Dial Indicators

Dial indicators are precision instruments used to measure the difference in size or location between a workpiece and a reference standard. Although they are capable of linear measurement, dial indicators are generally used for comparison measurements, such as checking the alignment and concentricity of a workpiece in a lathe. Here the comparison is made between the axis of the lathe and the axis of the part. Dial indicators are also used to align milling machine vises with the machine table by comparing the location of the vise jaw at each end of the vise. To perform an inspection, the dial indicator is set with a standard and, using the indicator, each workpiece is compared to the standard. Any variation from the preset size is then easily detected by reading the dial graduations. To use a dial indicator properly, you should first become familiar with how this instrument is constructed.

CONSTRUCTION OF A DIAL INDICATOR

As a machinist, you will be required to know how to read and use several types of dial indicators. As shown in Figure 7-1, the dial indicator resembles a clock. The *case* houses the geared indicating mechanism, the *dial*, and the *indicating hand* (pointer). Around the outside of the dial are the *bezel,* used to position

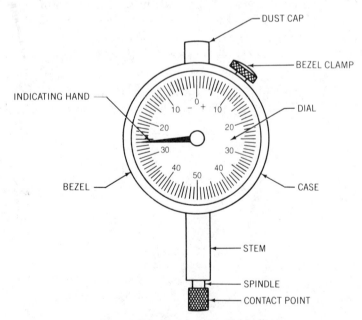

Figure 7-1 Parts of the dial indicator.

the zero on the dial in any position around the face of the indicator, and the *bezel clamp*, used to lock the position of the bezel. Below the case is the *stem*, which houses the *spindle*. On the end of the spindle is the *contact point*. The *dust cap*, on the top side of the case, is used to keep dust and dirt out of the indicator and also is a positive stop for the spindle.

READING A DIAL INDICATOR

Dial indicators are made in a wide variety of types and styles to suit their many applications. Generally, dial indicators are classified by their dial graduations, their range, or both. Most dial indicators used in the machine shop are graduated in increments of 0.001, 0.0002, and 0.0005 in. and have ranges of measurement from 0.025 to 6.000 in. Metric dial indicators usually have graduations of 0.01 and 0.002 mm in ranges of 0.5 to 150 mm.

The dials on most dial indicators have either *balanced* or *continuous* graduations (Figure 7-2). Balanced dials are generally used for comparative work and are graduated from zero in plus (+) or minus (–) units on each side of the dial. Continuous dials are usually used for linear measurements and are graduated in continuous units around the dial. Both types of dial indicator can be used for either measurement or comparison. The different graduations are used to make the dial indicators easier to read and harder to misinterpret.

Another feature often found on dial indicators that have larger ranges is the revolution counter (Figure 7-3). The revolution counter is used to show the number of complete revolutions made by the indicating hand. The numbers on the revolution counter sometimes refer to a specific distance. For example, the graduations on the revolution counter shown are 0 to 9. Since the range of the dial indicator is 1.000 in. and the graduations around the dial equal 0.100 in., each number on the revolution counter is equal to 0.100 in.

When using a dial indicator, you should first make note of the discrimination and range of the indicator. This information is usually found on the face of the dial (Figure 7-4). Next, note the direction in which the indicating hand travels. Dial indicators can be read in both the clockwise and counterclockwise directions. Generally, dial indicators read positive (+) in the clockwise direction and negative (–) in the counterclockwise direction (Figure 7-5). Finally, to read the measurement, simply read the position of the indicating hand and, if the dial indicator has a revolution counter, add this value to the dial reading (Figure 7-6). If the dial indicator does not have a revolution counter, only the dial

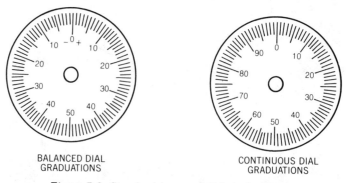

BALANCED DIAL GRADUATIONS   CONTINUOUS DIAL GRADUATIONS

Figure 7-2 Standard types of dial graduations.

Unit 7 Dial Indicators 83

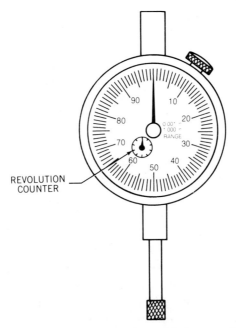

Figure 7-3 The revolution counter.

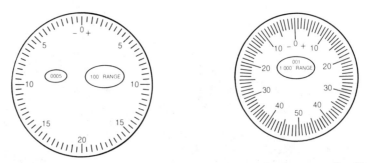

Figure 7-4 The discrimination and range are usually shown on the dial.

Figure 7-5 Negative and positive readings.

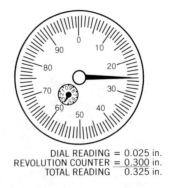

DIAL READING = 0.025 in.
REVOLUTION COUNTER = 0.300 in.
TOTAL READING   0.325 in.

Figure 7-6 Reading a dial indicator.

SECTION IV INDICATORS AND GAGES

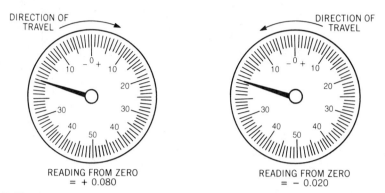

Figure 7-7 The direction of travel of the indicating hand has a direct effect on the indicator reading.

graduations are read. The exception to this is when the indicator has a range greater than one dial revolution but does not have a revolution counter. In these cases you will have to count the revolutions and add the value to the dial reading. Always pay close attention to the direction of movement of the indicating hand. As shown in Figure 7-7, this has a direct bearing on the indicator reading. Now that you know how to read a dial indicator, let's try a few practice problems (see Figure 7-8).

Practice Problem 1

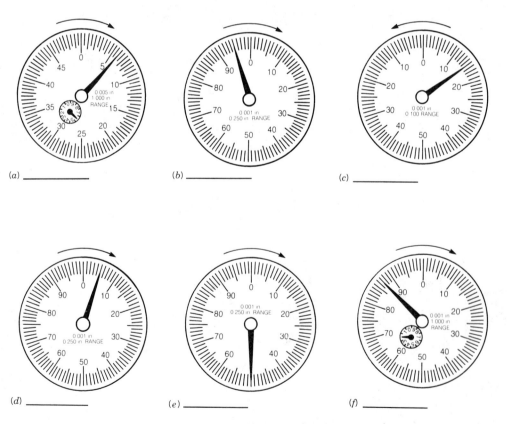

Figure 7-8 Practice problem 1.

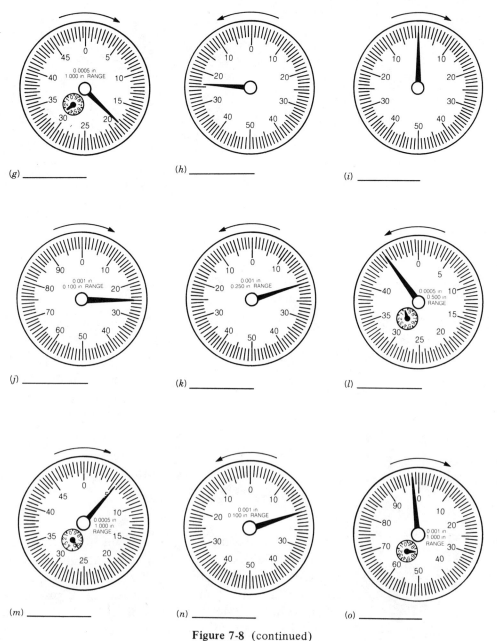

Figure 7-8 (continued)

Reading a metric dial indicator is similar to reading the inch-type indicators. First, determine the discrimination and range of the indicator. Next, note the direction of movement of the indicating hand. Finally, make the measurement by adding the value shown on the dial to the value shown by the revolution counter. If the dial indicator does not have a revolution counter, you will have to count the number of complete revolutions. In the cases where the indicator does not make a full revolution, only the value shown on the dial is considered in the measurement (Figure 7-9). Now that you can read a metric dial indicator, let's try a few practice problems (see Figure 7-10).

SECTION IV INDICATORS AND GAGES

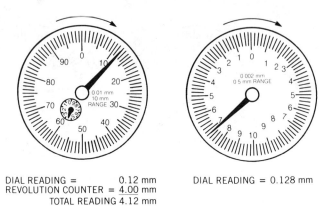

DIAL READING = 0.12 mm
REVOLUTION COUNTER = 4.00 mm
TOTAL READING 4.12 mm

DIAL READING = 0.128 mm

Figure 7-9 Reading a metric dial indicator.

Practice Problem 2

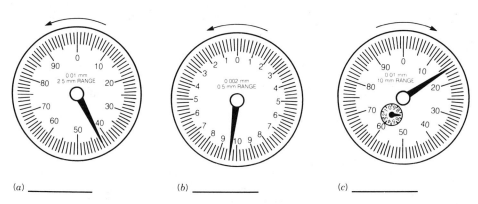

(a) _____   (b) _____   (c) _____

Figure 7-10 Practice problem 2.

USING A DIAL INDICATOR

When using a dial indicator, as shown in Figure 7-11, position the indicator so it can make the desired measurement. The indicator or the workpiece must be mounted so that the indicating hand is free to travel. Anything that would restrict its movement during a measurement must be changed. Always take time to insure the indicator will not strike anything during the measurement. This could affect the measurement and could also damage the indicator.

Dial indicators are generally used with a variety of attachments for mounting and locating the indicator. The attachments include the magnetic base, test stand, tool post holder, post assembly, and "G" clamp. There are also a wide range of indicator accessories that extend the measuring capabilities of the standard dial indicator. These include the deep hole attachment, the right angle attachment, and a wide selection of contact points (Figure 7-12). Figure 7-13 shows two typical applications where dial indicators are used to check alignment.

Once the indicator is properly positioned, bring the indicator into contact with the workpiece. Be careful not to strike the workpiece with the indicator and make sure the amount of movement you expect the indicator to make is within its range. Failing to do this could result in serious damage to the indicator. Once contact is made with the workpiece, you are ready to *zero* the

Unit 7 Dial Indicators 87

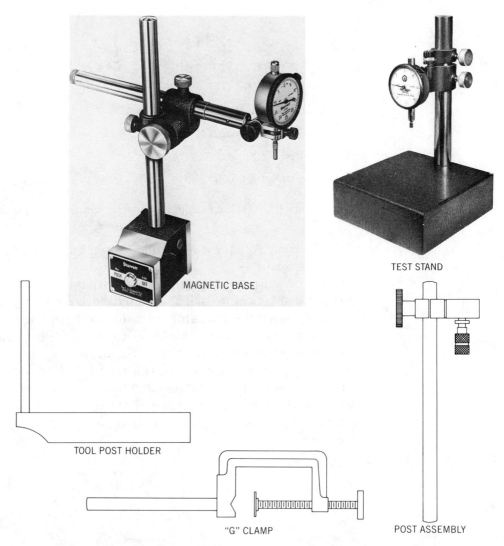

Figure 7-11 Holding devices for dial indicators. (Courtesy of L.S. Starrett Co.)

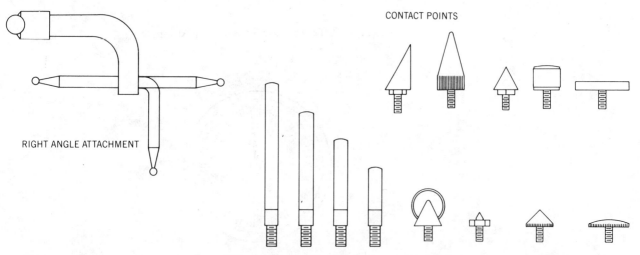

Figure 7-12 Dial indicator attachments.

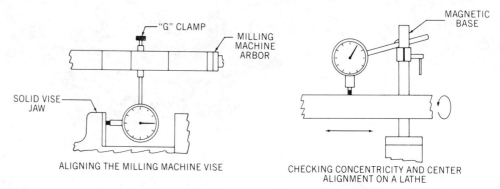

Figure 7-13 Typical examples of how indicators are used in the machine shop.

dial. The spindle or pivot of the indicator must be perpendicular to the surface to be measured.

When the indicator is not in contact with the workpiece, the indicating hand on most dial indicators is at the 9:00 o'clock position (Figure 7-14). To insure accuracy, move the indicator into the workpiece until the indicating hand passes the zero graduation (Figure 7-15). In cases where the indicator is used to show both positive and negative values, depress the spindle one-half to one full revolution of the indicating hand (Figure 7-16). With the indicating hand in this position, loosen the bezel clamp and turn the bezel until the zero graduation is aligned under the indicating hand. Tighten the bezel clamp and the indicator is *zeroed*. Any required measurements or comparisons can then be made with precision.

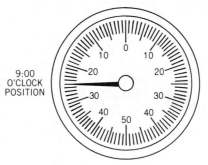

Figure 7-14 Indicating hand rest position.

Figure 7-15 Move indicating hand slightly past zero when zeroing the indicator.

Unit 7 Dial Indicators 89

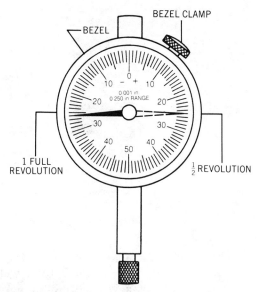

Figure 7-16 Indicating hand should be moved $\frac{1}{2}$ to 1 revolution for indicating positive and negative values.

VARIATIONS OF THE DIAL INDICATOR

Dial indicators are used for a wide variety of measuring applications. There are several variations of the basic dial indicator, including the test indicator, the back plunger indicator, the dial depth gage, the dial bore gage, and the dial thickness gage.

TEST INDICATOR

The test indicator (Figure 7-17) is widely used throughout the machine shop

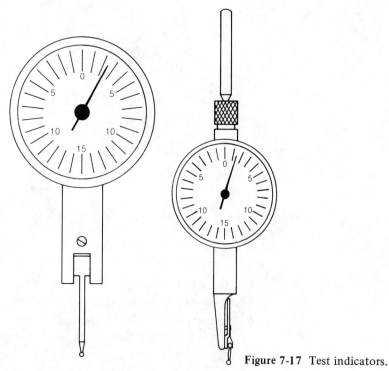

Figure 7-17 Test indicators.

SECTION IV INDICATORS AND GAGES

Figure 7-18 Using a test indicator in a vertical milling machine. (Courtesy of DoAll Co.)

for operations such as aligning, locating, and machine setups. As shown in Figure 7-18, this type of indicator is well suited for work in confined areas such as the spindle of a vertical milling machine. The advantage in using this type of indicator is its small size and adaptability. Test indicators, when used with their wide range of attachments, can easily fit into areas where it would be impossible to use a standard dial indicator.

BACK PLUNGER INDICATOR

The back plunger indicator (Figure 7-19) is made with the stem and spindle at a right angle to the mounting rod. Figure 7-20 shows a typical application of this type of dial indicator.

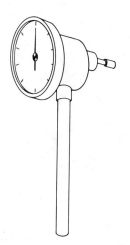
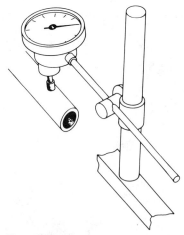

Figure 7-19 Bad plunger-type indicators.

Figure 7-20 Using a back plunger indicator in a lathe.

DIAL DEPTH GAGE

The dial depth gage (Figure 7-21) is used for measuring the depth of holes, slots, and grooves. Dial depth gages are available in a wide range of styles with accessories to meet almost any requirement. Figure 7-22 shows one type of dial depth gage with a knife edge base, conical contact point, and a lifting lever. Figure 7-23 shows the interchangable bases that can be used with these indicators.

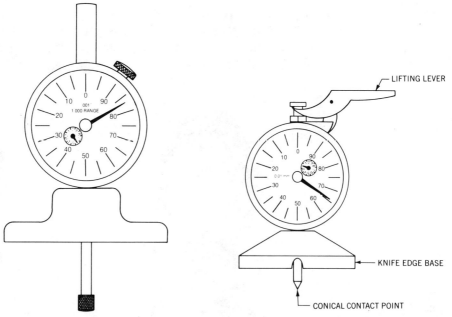

Figure 7-21 Dial depth gage.

Figure 7-22 Dial depth gage with special features.

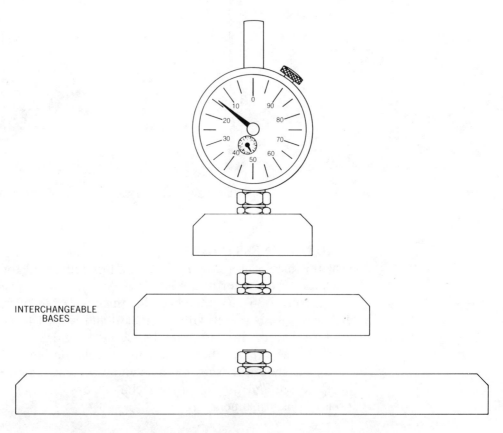

Figure 7-23 Interchangeable bases for dial depth gages.

DIAL BORE GAGES

The dial bore gage (Figure 7-24) is primarily used for checking the size and shape of round holes. These gages are available in sizes ranging from 0.037 to 12 in. (1 to 300 mm) in several styles and types.

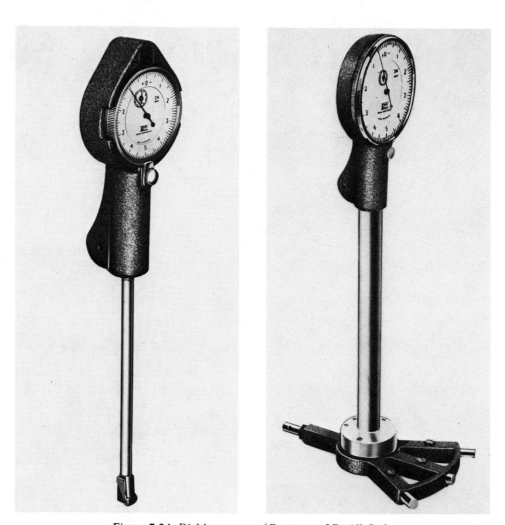

Figure 7-24 Dial bore gages. (Courtesy of DoAll Co.)

DIAL THICKNESS GAGES

Dial thickness gages (Figure 7-25) are primarily used for measuring either the thickness or diameter of a workpiece. Another similar variation is the *dial snap gage* (Figure 7-26), used to check the thickness or diameter of a workpiece. The dial snap gage is preset with a standard and is used to compare the size of the workpiece to the preset standard size; the dial thickness gage is actually used to measure a workpiece. Both gages are available in many sizes, types, and styles.

The dial indicators shown in this unit represent the majority of those you are likely to see and use in most machine shops, although there are other variations of this instrument.

Unit 7 Dial Indicators 93

Figure 7-25 Dial thickness gage. (Courtesy of L.S. Starrett Co.)

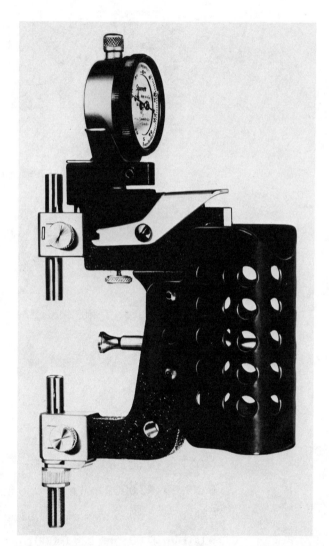

Figure 7-26 Dial snap gage. (Courtesy of L.S. Starrett Co.)

SECTION IV INDICATORS AND GAGES

SELF TEST

1. Identify the parts labeled A to J of the dial indicator shown in Figure 7-27.

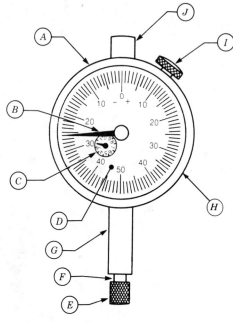

Figure 7-27

2. What is the discrimination of the indicator shown in problem 1? (Circle one.)

   a. 0.0005 in.

   b. 0.500 in.

   c. 5.000 in.

   d. 1.000 in.

3. What is the range of the indicator shown in problem 1? (Circle one.)

   a. 5.000 in.

   b. 0.500 in.

   c. 0.250 in.

   d. 0.0005 in.

4. How are the most inch-type dial indicators graduated? (Circle one.)

   a. 0.001 in., 0.002 in., and 0.005 in.

   b. 0.0001 in., 0.0002 in., and 0.050 in.

   c. 0.001 in., 0.0005 in., and 0.0001 in.

   d. 0.001 in., 0.005 in., and 0.0001 in.

5. How are most metric dial indicators graduated? (Circle one.)

   a. 0.0001 mm and 0.0001 mm

b. 0.01 mm and 0.002 mm

c. 0.1 mm and 0.2 mm

d. 0.001 mm and 0.12 mm

6. What types of dials are generally found on dial indicators? (Circle one.)

   a. Balanced and unbalanced

   b. Continuous and balanced

   c. Unbalanced and continuous

   d. Continuous and uncontinuous

7. In which direction must the indicating hand move to show positive readings?

8. In which direction must the indicating hand move to show negative readings?

9. List four standard holding devices used with dial indicators.

10. List three attachments that extend the measuring capabilities of the standard dial indicator.

11. List six variations of the dial indicator commonly found in the machine shop.

12. Make the readings for Figure 7-28 and write the correct answers in the spaces provided.

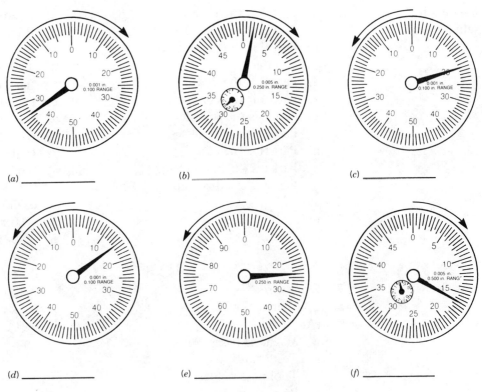

Figure 7-28

SECTION IV INDICATORS AND GAGES

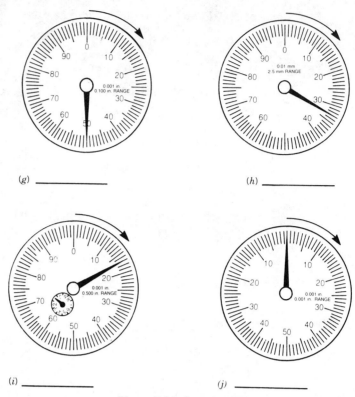

Figure 7-28 (continued)

Answers to Self Test

1. a. Part A is the bezel
   b. Part B is the indicating hand
   c. Part C is the revolution counter
   d. Part D is the dial
   e. Part E is the contact point
   f. Part F is the spindle
   g. Part G is the stem
   h. Part H is the case
   i. Part I is the bezel clamp
   j. Part J is the dust cap
2. a
3. b
4. c
5. b
6. b
7. Clockwise
8. Counterclockwise
9. (Any four)
   a. Magnetic bases
   b. Test stands
   c. Tool post holder
   d. Post assembly
   e. "G" clamp
10. a. Deep hole attachment
    b. Right angle attachment
    c. Contact points
11. a. Test indicator
    b. Back plunger indicator
    c. Dial depth gage
    d. Dial bore gage
    e. Dial thickness gage
    f. Dial snap gage
12. a. 0.066 in.
    b. 0.1515 in.
    c. 0.080 in.
    d. −0.080 in.
    e. −0.085 in.
    f. −0.76 mm
    g. 0.050 in.
    h. 0.33 mm
    i. 0.117 in.
    j. 0.100 in. or 0

Answers to Practice Problems

1. a. 0.606 in.
   b. 0.095 in.
   c. −0.085 in.
   d. 0.005 in.
   e. 0.050 in.
   f. 0.287 in.
   g. 0.3185 in.
   h. −0.024 in.
   i. 0.100 in., or 0
   j. 0.025 in.
   k. −0.080 in.
   l. −0.005 in.
   m. 0.5055 in.
   n. −0.080 in.
   o. 0.798 in.

2. a. −0.58 mm
   b. −0.096 mm
   c. 7.16 mm

UNIT 8

Gage Blocks and Fixed Gages

Gage blocks and fixed gages are precision gaging tools that have preset, or fixed, sizes. These tools are used for precision gaging operations and for calibrating other measuring tools.

GAGE BLOCKS

Modern standards for linear measurement are expressed in terms of wavelengths. The international inch is defined as equal to 41,929.399 wavelengths of Krypton 86 orange-red radiation. Since it would be impossible to use these wavelengths to measure a workpiece in the shop, a standard with physical form had to be developed. Gage blocks provide a means by which this wavelength standard can be used for practical measurement.

Gage blocks are machined, ground, and lapped to a very high degree of accuracy in size, flatness, and parallelism. These blocks are the accepted standard for linear measurement in industry.

TYPES OF GAGE BLOCKS

Gage blocks are made from a variety of materials and in three basic shapes: square, rectangular, and cylindrical (Figure 8-1). They are generally made of hardened and stabilized tool steel, stainless steel, or tungsten carbide.

Gage blocks are normally made in sets that consist of several blocks of varying sizes. The blocks can be used individually or together to obtain almost any size within the range of the set. A typical set of 81 gage blocks (Figure 8-2)

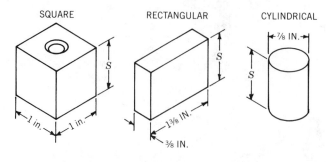

S = GAGE BLOCK SIZE

Figure 8-1 Typical shapes of gage blocks.

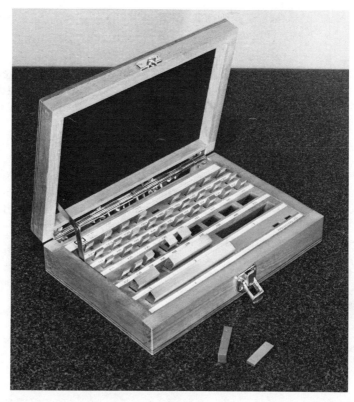

Figure 8-2 Gage block set. (Courtesy of Federal Products Corp.)

is capable of producing well over 100,000 different sizes. The normal range of sizes for gage blocks is between 0.010 and 20.000 in. for inch-type blocks and 0.20 and 500.00 mm for metric blocks.

GAGE BLOCK ACCURACY

Gage blocks are made in four grades of accuracy, 0.5, 1, 2, and 3 (formally AAA, AA, A+, and A). The exact tolerance limits for each of these grades is shown in Figure 8-3.

Since absolute perfection is almost impossible to achieve (and, if achieved, impossible to measure), these tolerance values have been established to control the sizes of the blocks within each grade. The exact size of each block within a

FEDERAL ACCURACY GRADES

ACCURACY GRADE	FORMER DESIGNATION	TOLERANCE	
		ENGLISH SYSTEM (inch)	METRIC SYSTEM (millimeter)
0.5	AAA	±0.000001	±0.00003
1	AA	±0.000002	±0.00005
2	A+	+0.000004 −0.000002	+0.0001 −0.00005
3	A	+0.000006 −0.000002	+0.00015 −0.00005

Figure 8-3 Accuracy grades of gage blocks. (Courtesy of Federal Products Corp.)

set is generally shown in the certificate of calibration that comes with each set. This calibration of gage blocks is normally traceable to the U.S. National Bureau of Standards in Washington, D.C.

The nominal size of each gage block is generally etched on the side of the block along with the name of the manufacturer and the block serial number. This serial number identifies the set to which the gage block belongs.

USING GAGE BLOCKS

Gage blocks are used for two general purposes, checking and calibrating other measuring tools such as micrometers and verniers, and gaging the sizes of workpieces (Figure 8-4). In either case, gage blocks are normally used in groups of two or more to equal the desired size. Gage blocks are held together by *wringing* the blocks (Figure 8-5).

$\frac{1}{8}$ in. OVERLAP

PRESS LIGHTLY AND SLIDE BLOCKS TOGETHER

SLIDE TOGETHER UNTIL COMPLETELY MATED

CHECK TO INSURE THEY WILL ADHERE TO EACH OTHER

Figure 8-6 Wringing gage blocks.

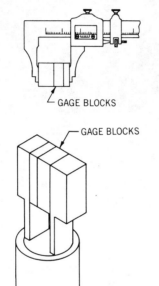

Figure 8-4 Typical uses of gage blocks.

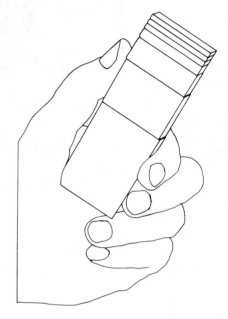

Figure 8-5 Gage block stack.

Wringing is a process of sliding one block on another for the purpose of building a gage block stack. Gage blocks, when wrung, are held together by a combination of the attraction of one block to another and by a minute film of oil or moisture between the blocks reacting to the almost perfectly flat surface.

Wringing gage blocks is basically a five-step procedure (Figure 8-6). The first step is cleaning. Gage blocks must be free of dirt, oil, or grit to adhere properly to each other. Next, the edges of the blocks should be overlapped approximately $\frac{1}{8}$ in. Then, applying slight downward pressure, begin to slide the blocks together. Once the gaging surfaces are completely mated, check the stack to be sure the blocks are properly wrung together. If you follow this process, you should be able to gage blocks properly.

Unit 8 Gage Blocks and Fixed Gages 101

CALCULATING GAGE BLOCK STACK HEIGHT

When using gage blocks to check a certain size, be sure the stack height is exactly the same as the dimension to be checked. The easiest way to calculate the proper stack height is through subtraction. To simplify the process of selecting the proper blocks for a certain size, eliminate the last decimal value in each step of the calculation, as shown here. Always use the fewest blocks possible. For example, to check a dimension of 3.8672, you should select the following from the blocks listed in Figure 8-7.

$$\begin{array}{r} \text{Required size} = 3.8672 \text{ in.} \\ \text{The 0.0002 is eliminated with a } \underline{0.1002} \text{ block} \\ 3.7670 \\ \text{The 0.007 is eliminated with a } \underline{0.1170} \text{ block} \\ 3.6500 \\ \text{The 0.6500 is eliminated with a } \underline{0.6500} \text{ block} \\ 3.0000 \\ \text{Finally, the 3.0000 is eliminated with a } \underline{3.0000} \text{ block} \\ 0.0000 \end{array}$$

The blocks you should select for this stack are: 1 – 0.1002 in., 1 – 0.1170 in., 1 – 0.6500 in., and 1 – 3.0000 in. blocks. Now that you know how to calculate gage block stack heights, select the proper blocks for the following sizes (see

Inch Gage Blocks

Size in Inches

0.0001-in. steps	0.1001	0.1002	0.1003	0.1004	0.1005	0.1006	0.1007	0.1008	0.1009	
	0.101	0.102	0.103	0.104	0.105	0.106	0.107	0.108	0.109	
	0.110	0.111	0.112	0.113	0.114	0.115	0.116	0.117	0.118	0.119
0.001-in. steps	0.120	0.121	0.122	0.123	0.124	0.125	0.126	0.127	0.128	0.129
	0.130	0.131	0.132	0.133	0.134	0.135	0.136	0.137	0.138	0.139
	0.140	0.141	0.142	0.143	0.144	0.145	0.146	0.147	0.148	0.149
0.050-in. steps	0.050	0.100	0.150	0.200	0.250	0.300	0.350	0.400	0.450	
	0.500	0.550	0.600	0.650	0.700	0.750	0.800	0.850	0.900	0.950
1.000-in. steps	1.000	2.000	3.000	4.000						

Metric Gage Blocks

Size in Millimeters

0.002-mm steps	1.001	1.002	1.003	1.004	1.005	1.006	1.007	1.008	1.009	
	1.01	1.02	1.03	1.04	1.05	1.06	1.07	1.08	1.09	
	1.10	1.11	1.12	1.13	1.14	1.15	1.16	1.17	1.18	1.19
0.01-mm steps	1.20	1.21	1.22	1.23	1.24	1.25	1.26	1.27	1.28	1.29
	1.30	1.31	1.32	1.33	1.34	1.35	1.36	1.37	1.38	1.39
	1.40	1.41	1.42	1.43	1.44	1.45	1.46	1.47	1.48	1.49
0.5-mm steps	0.5	1.0	1.5	2.0	2.5	3.0	3.5	4.0	4.5	5.0
	5.5	6.0	6.5	7.0	7.5	8.0	8.5	9.0	9.5	
10.0-m steps	10	20	30	40	50	60	70	80	90	100

Figure 8-7 Standard gage block sizes.

102 SECTION IV INDICATORS AND GAGES

DIMENSIONS	GAGE BLOCK SIZES					NUMBER OF GAGE BLOCKS
(a) 4.6168 in.						
(b) 02.4011 in.						
(c) 0.6679 in.						
(d) 0.745 in.						
(e) 1.634 in.						
(f) 0.3333 in.						
(g) 42.414 mm.						
(h) 52.007 mm.						
(i) 75.288 mm.						

Figure 8-8 Practice problem 1.

Figure 8-8) remembering to use both the correct size blocks and the fewest blocks possible.

GAGE BLOCK ACCESSORIES

Gage blocks have a wide assortment of accessories, generally furnished in sets, that greatly extend the use of these tools. Gage block accessories include scriber points, center points, trammel points, cylindrical jaws, flat jaws, block holders, and base blocks (Figure 8-9). Each of these accessories is precision lapped to the same accuracy as the gage blocks with which they are used. These

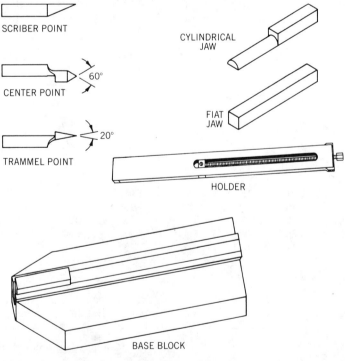

Figure 8-9 Gage block accessories.

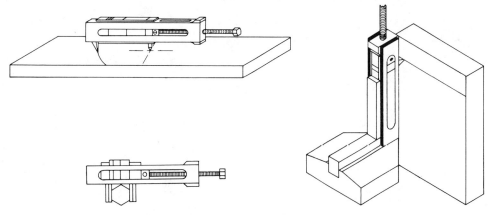

Figure 8-10 Typical applications of gage blocks.

accessories are wrung together with the blocks and are used either as a stack or in the holder. Figure 8-10 shows how gage blocks and their accessories are used together.

CARE OF GAGE BLOCKS

1. Make sure the blocks are completely clean before wringing. Do not wring dirty gage blocks.

2. After using gage blocks, take the stack apart and put the blocks in their case; keep them there when they are not in use.

3. Keep gage blocks coated with a thin film of oil or corrosion-resistant compound.

4. Do not treat gage blocks roughly. Look for damage and notify the person responsible for gage block care if burrs or nicks are found.

FIXED GAGES

Fixed gages are used to compare the size of a part feature to a preset standard. Gages cannot measure; they can only compare. A gage can tell you if a part is correct or incorrect but cannot tell you by how much a part is incorrect. Fixed gages are made in many styles and sizes but can be grouped into three general categories: plug gages, ring gages, and snap gages.

Some fixed gages are designed as single-element gages, but most are double-element gages. In other words, these gages have either one or two gaging surfaces. Most double-element fixed gages are also called GO-NO GO gages.

THE GO-NO GO METHOD OF GAGING

GO and NO GO gages are designed to check the sizes of a part feature at its upper and lower limits of size. As shown in Figure 8-11, the GO end of the gage is made to the smallest allowable size of the hole. The NO GO element is made to the largest allowable size of the hole. If the hole is within tolerance, the GO gage should fit into the hole, but the NO GO gage should be too big to fit (Figure 8-12). Gages for external dimensions use this same principle, the

SECTION IV INDICATORS AND GAGES

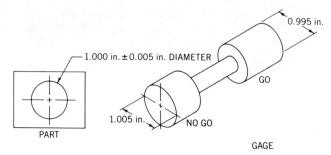

Figure 8-11 Plug gage for checking upper and lower limits of hole.

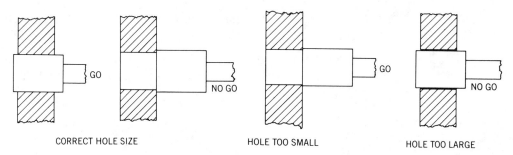

Figure 8-12 Using a plug gage with GO and NO GO gaging elements.

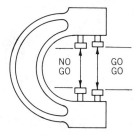

Figure 8-13 GO and NO GO gage for external features (snap gage).

only difference is that the GO and NO GO functions are reversed. For external dimensions the GO gage is made to the largest size of the dimensions, and the NO GO gage is made to the smallest size (Figure 8-13).

PLUG GAGES

Plug gages (Figure 8-14) are available in three basic styles, cylindrical, thread, and taper.

Cylindrical plug gages are used to check the size and form of round holes. These gages are made with either double-end or progressive gaging elements (Figure 8-15). The double-end style has the GO gage at one end and the NO GO gage at the other end. The progressive style has both the GO and NO GO on one end.

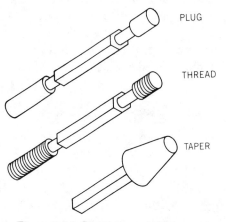

Figure 8-14 Plug gage variations.

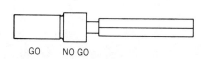

Figure 8-15 Progressive plug gage.

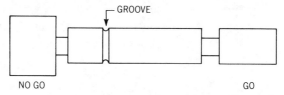

Figure 8-16 Marking the NO GO end of a plug gage.

In most cases the GO and NO GO gaging ends are identified by either a color system, such as red (NO GO) and green (GO), or by a groove cut in the handle at the NO GO end (Figure 8-16). On most plug gages, the GO end is normally longer than the NO GO end of the gage.

Threaded plug gages are similar in design to cylindrical plug gages. The principal difference is the gaging elements. Threaded plug gages have gaging elements that are made to suit the pitch diameter and major diameter of the specific thread class they are intended to gage. For example, a $\frac{1}{2}$-13-2B thread has a pitch diameter range of 0.4500 to 0.4565 in. and a major diameter of 0.5000 in. The thread plug gage for this thread would have a GO gage with a pitch diameter of 0.4500 in. and a NO GO gage with a pitch diameter of 0.4565 in. and a major diameter of 0.5000 in. In order for a thread to pass inspection, it would have to accept the GO gage and reject the NO GO gage. If the GO gage cannot enter the hole, the pitch diameter is too small. If the NO GO gage can enter the hole, either the pitch diameter or the thread form is incorrect. In either case the thread would have to be rejected as out of tolerance.

Taper plug gages are used to check the size and form of tapered holes. A gage is first coated with an inklike or Prussian blue compound and inserted into the tapered hole. When it is removed, if the blueing shows a uniform pattern over its entire length or over the length of the tapered hole, the taper is correct. If more blue is transferred off either end, the taper is either too steep or not steep enough.

The diameter of the tapered hole is also checked by a plug gage. Some taper plug gages are made with lines on the gaging surface. To pass inspection, the large end of the taper must lie between these lines when the gage is firmly seated in the part. Other variations of the taper plug gage use other methods to determine the proper diameter, such as flats, grooves, or simply making the diameter at the ends of the plug gage equal to the maximum and minimum sizes of the tapered hole.

RING GAGES

Ring gages (Figure 8-17), like plug gages, are also available in three general

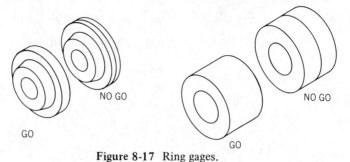

Figure 8-17 Ring gages.

types, cylindrical, thread, and taper. These gages are generally made in sets of one GO gage and one NO GO gage. The use of these gages is similar to that of the plug gage except that ring gages are used to check external part features.

SNAP GAGES

Snap gages (Figure 8-18) are used for gaging external surfaces or diameters. Snap gages can also be used for inspecting external threads (Figure 8-19). These gages are generally made so that the distance between the gaging elements can be adjusted. Snap gages are also designed to check both the GO and NO GO dimensions. Usually, the front set of gaging elements is used for the GO dimension and the rear set for the NO GO size.

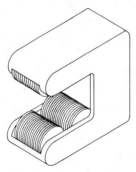

Figure 8-18 Snap gage for checking external threads.

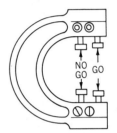

Figure 8-19 Location of GO and NO GO elements on a snap gage.

CARE OF FIXED GAGES

Fixed gages are precision tools and, to maintain their accuracy, must be treated properly.

1. Handle gages carefully and avoid dropping, nicking, or scratching.

2. When you are finished with a fixed gage, coat it with a corrosion-resistant compound and put it back in its case.

3. Never force a fixed gage. Forcing will score the workpiece or gage.

4. Treat fixed gages as carefully as you would gage blocks. There is a sizable investment in each of these gages: TREAT THEM PROPERLY.

SELF TEST

1. List the three basic shapes of gage blocks.

2. Which grade of gage block is the most accurate? (Circle one.)

    a. 0.5

    b. 1

    c. 2

    d. 3

3. Which of the following is *not* a function of gage blocks? (Circle one.)
   a. Checking the diameter of a hole
   b. Measuring the length of a bar stock
   c. Calibrating a vernier depth gage
   d. Checking an outside micrometer

4. How are gage blocks held together when in use? (Circle one.)
   a. Bolts
   b. Clamps
   c. Friction
   d. Wringing

5. Calculate the stack heights for the following dimensions. (Remember to use the least number of blocks possible.) Refer to the chart of gage block sizes (Figure 8-7) to answer these questions.
   a. 0.451 in.
   b. 6.157 mm
   c. 2.2332 in.
   d. 11.0001 in.

6. List seven typical accessories used with gage blocks.

7. Which of the following *cannot* be done with a fixed gage? (Circle one.)
   a. Measuring the diameter of a bar
   b. Checking the diameter of a hole
   c. Inspecting a tapered hole
   d. Checking the size of a thread
   e. All can be checked by fixed gages.

8. What are the three basic types of fixed gages?

9. How many gaging elements does a progressive plug gage have? (Circle one.)
   a. 1
   b. 2
   c. 3
   d. 4

10. What are the upper and lower gaging elements called? (Circle one.)
    a. Big and small
    b. OK and NOT OK
    c. GO and LOW
    d. GO and NO GO

Answers to Self Test

1. a. Square
   b. Rectangular
   c. Cylindrical
2. a
3. b
4. d
5. a. 0.100 in – 0.101 in.
      – 0.250 in.   (3 blocks)
   b. 1.007 mm – 151 mm
      – 4.0 mm   (3 blocks)
   c. 0.1002 in. – 0.133 in.
      – 2.000 in.   (3 blocks)
   d. 0.1001 in. – 0.900 in.
      – 1.000 in. – 2.000 in.
      – 3.000 in. – 4.000 in.
                           (6 blocks)
6. a. Scriber points
   b. Center points
   c. Trammel points
   d. Cylindrical jaws
   e. Flat jaws
   f. Block holders
   g. Base blocks
7. e
8. a. Plug
   b. Ring
   c. Snap
9. b
10. d

Answers to Practice Problem

1. a. 0.1008 in – 0.116 in.
      0.400 in. – 4.000 in.
                           (4 blocks)
   b. 0.1001 in. – 0.101 in.
      – 0.200 in. – 2.000 in.
                           (4 blocks)
   c. 0.1009 in. – 0.117 in.
      0.450 in.   (3 blocks)
   d. 0.145 in. – 0.600 in.
                           (2 blocks)
   e. 0.134 in. – 0.500 in.
      – 1.000 in.   (3 blocks)
   f. 0.1003 in. – 0.103 in.
      – 0.130 in.   (3 blocks)
   g. 1.004 mm – 1.410 mm
      – 40.00 mm   (3 blocks)
   h. 1.007 mm – 1.00 mm
      – 50.00 mm   (3 blocks)
   i. 1.008 mm – 1.280 mm
      – 3.00 mm – 70.00 mm
                           (4 blocks)

UNIT 9

Comparison Gages

Comparison gages are semiprecision gages used to inspect the conformity of a part to a standard shape or size. They are divided into two general categories, size gages and form gages.

SIZE GAGES

Size gages are used to check the specific size of an item. Generally, these gages are made to conform to an established standard such as a gage size or a linear or angular value.

DRILL GAGES

Drill gages (Figure 9-1) are used to check the sizes of twist drills that have had their size marking worn off. These gages provide a quick way to check the sizes of twist drills. When extreme accuracy is important, the drill should be checked with a micrometer; however, for general sizing, these gages work well.

Drill gages are available in four basic sizes, fractional ($\frac{1}{16}$ to $\frac{1}{2}$), letter (A to Z), number (1 to 60), and small number (61 to 80). Metric models are also available to check the size of millimeter drills. In addition to checking twist drills, these gages check drill blanks and wire sizes.

Figure 9-1 Drill gage. (Courtesy of L.S. Starrett Co.)

SHEET, PLATE, AND WIRE GAGES

Sheet, plate, and wire gages (Figure 9-2) are used to check and identify the wide range of sheet metal, plate, and wire sizes. Since there are so many different scales of gage sizes, no one gage can check every type of material. There are at least six different standard scales used for sheet, plate, and wire, and each has a different value for each gage number. Before using one of these gages, identify the gaging system originally used for sizing the material.

Figure 9-2 Sheet, plate, and wire gage. (Courtesy of L.S. Starrett Co.)

THICKNESS GAGES

The thickness or feeler gage (Figure 9-3) is used for checking clearances, gaging narrow slots, and other similar applications. This gage consists of a frame that contains a variety of thin metal leaves, generally ranging in size from 0.0015 to 0.035 in. (0.03 to 3.00 mm).

These leaves can be used individually or in combination with each other to obtain a certain size. The most familiar application of the thickness gage is in setting the gap in automobile spark plugs and distributor points.

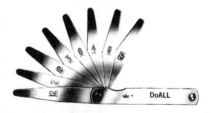

Figure 9-3 Thickness gage. (Courtesy of DoAll Co.)

ANGLE GAGES

The angle gage (Figure 9-4) is used for checking angular part features. This gage, similar in construction to the thickness gage, has a frame and a series of different size leaves. Angle gages are available with ranges of 1 to 45°. Figure 9-5 shows an appropriate use of this type of gage.

Unit 9   Comparison Gages   111

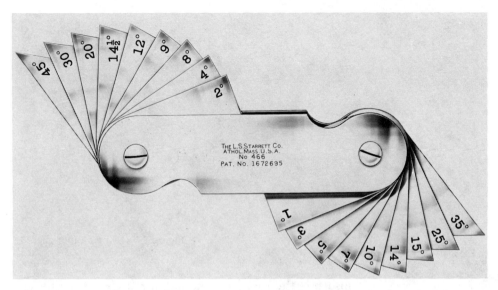

Figure 9-4 Angle gage. (Courtesy of L.S. Starrett Co.)

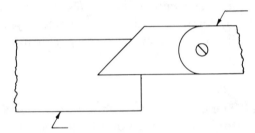

Figure 9-5 Typical use of angle gage.

TAPER GAGES

Taper gages (Figure 9-6) are a combination of a gage and a rule. These gages are used for checking the diameter of holes and the widths of slots and grooves by

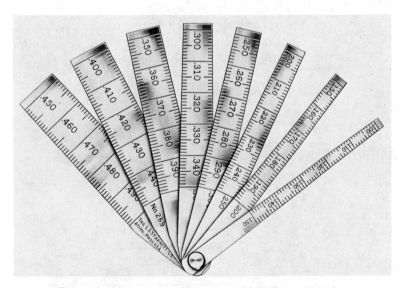

Figure 9-6 Taper gage. (Courtesy of L.S. Starrett Co.)

SECTION IV INDICATORS AND GAGES

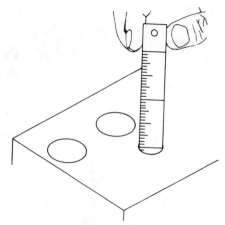

Figure 9-7 Using a taper gage.

inserting the gage into the hole or slot until it stops. The reading is then taken from the scale on the face of the gage blade. These gages are available in both inch and millimeter models. The inch type has either fractional or decimal inch graduations, and the millimeter type is graduated in 0.02-mm increments. Figure 9-7 shows how this gage is used to measure the diameter of a hole.

EDGE FINDERS

The edge finder (Figure 9-8), although not actually a gage, is a valuable setup tool with which you should be acquainted. The edge finder is used to align the spindle of a machine tool to the workpiece. Two basic styles of edge finders are the edge locating type and the line locating type. Some edge finders have both types on the same tool (Figure 9-9). The edge finder is placed in the spindle of the machine and, as the spindle slowly rotates, is brought into contact with the workpiece. When the end and body of the edge finder are aligned, the edge finder will "jump" off center. This indicates that the edge is located. In the case of the line-type edge finder, when the tool "jumps," the line and spindle center are aligned. However, in the case of the edge type, you must either add or subtract half the diameter of the edge finder to locate the spindle centerline. Figure 9-10 shows how this tool is used to locate an edge.

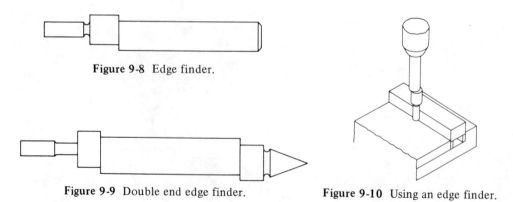

Figure 9-8 Edge finder.

Figure 9-9 Double end edge finder.

Figure 9-10 Using an edge finder.

Figure 9-11 Wiggler set. (Courtesy of DoAll Co.)

WIGGLERS

The wiggler (Figure 9-11) is a variation of the edge finder. The principal difference is the range of applications. The wiggler has a series of interchangeable contact points that are used to locate the spindle from a line, edge, or hole or with a test indicator.

TOOLMAKERS' BUTTONS

Toolmakers' buttons (Figure 9-12) are used to determine the exact location of holes or similar part features. These buttons are installed on the workpiece in the approximate location of the hole. The buttons are held in place with a small screw threaded into the workpiece. Once installed, the buttons are positioned exactly by using verniers, micrometers, or gage blocks. After the buttons have been properly positioned, the screws are tightened to lock the buttons in

Figure 9-12 Toolmakers' buttons. (Courtesy of L.S. Starrett Co.)

place. The workpiece is then put on the machine and the buttons are aligned with the machine spindle with a test indicator. Once located and aligned, the buttons are removed and the part feature is machined. As shown in Figure 9-12, one button is longer than the other three. This permits two buttons, one long and one short, to be used close together. The added height of the longer button allows it to be located without disturbing the other button. The accuracy and repeatability of modern machine tools is making these location methods obsolete.

FORM GAGES

Form gages are used to check the size and shape of part features or the distance from one point on a screw thread to the corresponding point on an adjacent thread.

SCREW PITCH GAGES

Screw pitch gages (Figure 9-13) are used to check the pitch of parts such as threaded shafts, bolts, and nuts. These gages consist of a frame and several leaves that represent the variety of standard thread pitches. The gage is placed on the threaded portion of the workpiece and the threads of the workpiece are compared to the thread pattern of the gage. Occasionally several leaves must be used until the one gage is found that fits the thread exactly. To find the pitch, simply read the number marked on each leaf.

Depending on the gage, each leaf is marked either with the number of threads per inch or the pitch of the thread in millimeters. Screw pitch gages are available in many styles and ranges of screw pitches; the normal range of these gages is 4 to 84 threads per inch, or 0.25 to 11.5 mm for the metric model.

Figure 9-13 Screw pitch gage. (Courtesy of DoAll Co.)

CENTER GAGES

Center gages (Figure 9-14) are used to check the 60° angle of threading tools. They also serve as templates for grinding threading tools and are used for referencing and aligning the threading tool when a lathe is set up for cutting threads. These applications are shown in Figure 9-15.

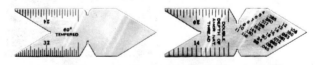

Figure 9-14 Center gage. (Courtesy of DoAll Co.)

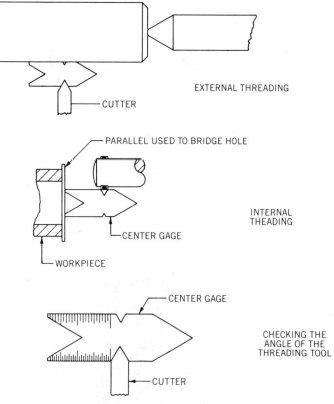

Figure 9-15 Typical uses of center gages.

29° SCREW THREAD GAGES

The 29° screw thread gage, or Acme thread gage (Figure 9-16), is used to check the angle of threading tools used to cut Acme threads. As with the center gage, this gage is also used as a grinding gage and a setup tool for cutting Acme threads on a lathe (Figure 9-17).

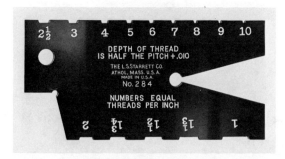

Figure 9-16 Twenty-nine degree screw thread gage. (Courtesy of L.S. Starrett Co.)

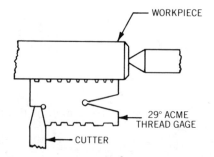

Figure 9-17 Using the 29° screw thread gage.

RADIUS GAGES

Radius gages are available in a wide selection of styles to suit their many uses. A few typical styles and uses are shown in Figure 9-18. The principal purpose of these gages is to check the form, size, and contour of radii. These gages are available in fractional inch, decimal inch, and millimeter sizes.

116 SECTION IV INDICATORS AND GAGES

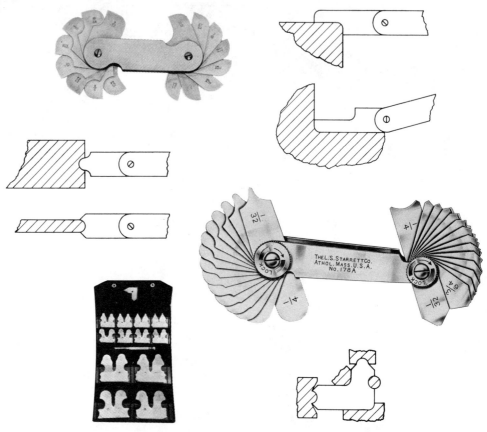

Figure 9-18 Radius gages and their uses. (Courtesy of L.S. Starrett Co.)

SELF TEST

1. How many basic type of drill gages are there for inch size drills? (Circle one.)

   a. 2

   b. 4

   c. 6

   d. 8

2. Which of the following is *not* a standard drill gage. (Circle one.)

   a. Number size (61 to 80)

   b. Fractional size ($\frac{1}{64}$ to $\frac{1}{16}$)

   c. Number size (1 to 60)

   d. Letter size (A to Z)

3. How many different gaging systems are used for sheets, plates, and wires? (Circle one.)

   a. 2

b. 4

c. 6

d. 8

4. What is the standard range of an angle gage? (Circle one.)

    a. 1 to 45°

    b. 1 to 30°

    c. 0 to 45°

    d. 0 to 90°

5. Which of the following is *not* a standard scale used for the taper gage? (Circle one.)

    a. 0.001 in.

    b. 0.02 mm

    c. $\frac{1}{64}$ in.

    d. 0.0001 cm

6. What is the purpose of an edge finder? (Circle one.)

    a. Locate holes

    b. Locate lines

    c. Reference holes to each other

    d. Reference the part to the spindle

7. Which of the following can be used to locate holes precisely? (Circle one.)

    a. Drill gage

    b. Feeler gage

    c. Toolmakers' buttons

    d. Machinists' buttons

8. What is the purpose of form gages? (Circle one.)

    a. Check contours and location of part features

    b. Check size and location of part features

    c. Check shape and size of part features

    d. Check location and form of part features

9. What is a screw pitch gage used for? (Circle one.)

    a. Check thread size

    b. Check thread pitch

    c. Check thread diameter

    d. Check pitch diameter

10. What is the angle of a center gage? (Circle one.)
    a. 45°
    b. 29°
    c. 60°
    d. $14\tfrac{1}{2}°$

11. What is another name for the 29° screw thread gage? (Circle one.)
    a. Buttruss thread gage
    b. Center gage
    c. Acme thread gage
    d. Pitch gage

12. Which of the following is *not* checked with a radius gage? (Circle one.)
    a. Size of radius
    b. Contour of radius
    c. Form of radius
    d. Location of radius

13. Which of the following is *not* a standard graduation used with radius gages? (Circle one.)
    a. $\tfrac{1}{8}$ mm
    b. 2 mm
    c. 0.100 in.
    d. $\tfrac{1}{8}$ in.

14. List three types of gages that are used to check sizes.

15. List three types of gages that are used to check forms.

Answers to Self Test

1. b
2. b
3. c
4. a
5. d
6. d
7. c
8. c
9. b
10. c
11. c
12. d
13. a
14. Any three of the following are correct.
    a. Drill gage
    b. Sheet, plate, and wire gages
    c. Thickness
    d. Angle gage
    e. Taper gage
15. Any three of the following are correct.
    a. Screw pitch gage
    b. Center gage
    c. 29° screw thread gage
    d. Radius gage

UNIT 10

Levels, Straightedges, and Precision Squares

Levels, straightedges, and squares are used for inspecting flat surfaces and right angles. Although these tools are not actually classified as gages, they do serve basically the same purpose.

LEVELS

Most levels used in the machine shop are the spirit, or bubble, type and are used in a wide range of workpiece setups and machine tool installations. A few of the more common levels include the pocket level, machinists' level, cross test level, and precision level.

POCKET LEVELS
The pocket level (Figure 10-1) is used for simple leveling tasks such as workpiece setups. This type of level is generally available in 2- and 3-in. lengths and is well suited for almost any noncritical leveling job.

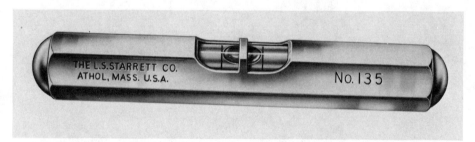

Figure 10-1 Pocket level. (Courtesy of L.S. Starrett Co.)

MACHINISTS' LEVELS
The machinists' level (Figure 10-2) is generally made in two styles, the plain level and the three-way model, and is well suited for semiprecision leveling applications. Unlike the pocket level, the machinists' level has an adjustable main vial and can be calibrated as necessary. The three-way type, in addition to an adjustable main vial, has a cross test and a plumb vial. The plumb vial is read when the level is used in a vertical position. This type of level is generally available in 4 and 6 in. sizes.

119

Figure 10-2 Machinists' levels. (Courtesy of L.S. Starrett Co.)

CROSS TEST LEVELS

Cross test levels (Figure 10-3) are generally used for leveling in two directions. This feature is especially useful for setting up workpieces in machines such as milling machines or drill presses. A useful variation of the cross test level is the bulls-eye level, with which a complete surface can be checked at a single setting. By design, the bubble in a bulls-eye level will automatically go to the high point of the surface being leveled.

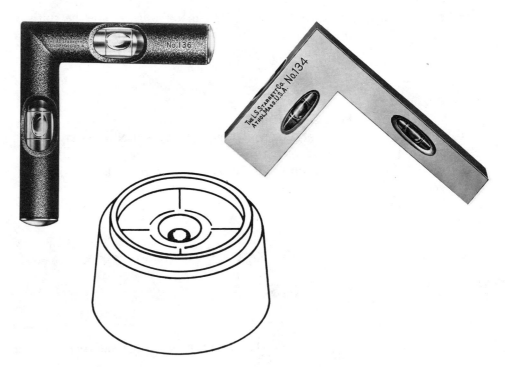

Figure 10-3 Cross test levels. (Courtesy of L.S. Starrett Co.)

PRECISION LEVELS

The precision level (Figure 10-4) is used for high-precision leveling applications such as the installation of machine tools. This level is generally made in lengths of 10 and 12 in. and has precision graduations on the vials. These graduations indicate the amount of variation from level in increments of 0.0005 in. per 12 in., or to an angular accuracy of 10 sec ($\frac{1}{360}$ of a degree).

Figure 10-4 Precision level. (Courtesy of L.S. Starrett Co.)

When not in use, the precision level should always be kept in its case to prevent it from being damaged and therefore help to insure its accuracy. Remember, this level is a precision tool and should always be treated as such. Never treat it roughly or abuse it.

STRAIGHT-EDGES

Straightedges (Figure 10-5) are used to inspect the conformity of flat surfaces to a straight plane. They are generally available in lengths of 12 to 72 in. and are useful for checking parts such as machine tables, automobile cylinder heads, or any other item that must have a flat surface. The straightedge is held on edge against the workpiece, and any variation from a true flatness can be detected by viewing the light visible between the part and the straightedge. To measure the amount of warp, or distortion, a feeler gage can be used. Straightedges are generally made from hardened tool steel with either straight or beveled edges.

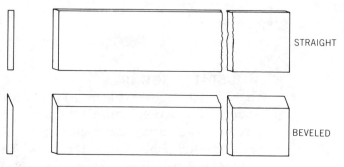

Figure 10-5 Straightedges.

PRECISION SQUARES

The precision square is a useful and accurate tool for inspecting right angles and perpendicularity. The precision square is available in many styles and types; the more common variations are the solid square, adjustable square, and cylindrical square.

SOLID SQUARES

The solid square (Figure 10-6) consists of a hardened and ground blade and beam that are pressed and pinned together at exactly 90°. This square can be used to check both internal and external right angles. Solid squares are available in sizes ranging from 2 to 12 in. with either straight or beveled blades.

When using this square, be sure it is properly positioned in relation to the workpiece (Figure 10-7). Keep the square positioned at a 90° angle to insure correct contact and make more accurate comparisons.

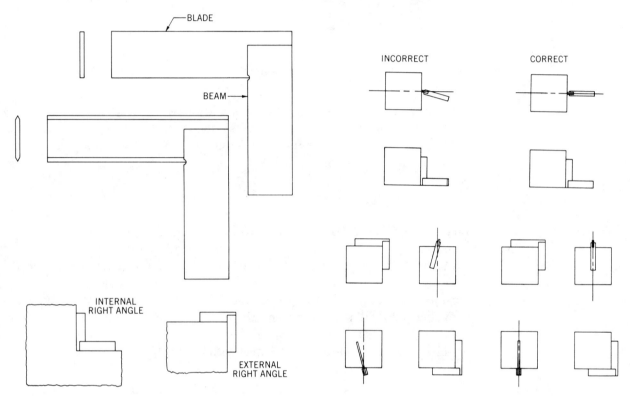

Figure 10-6 Solid squares.

Figure 10-7 Using a solid square.

ADJUSTABLE SQUARES

The adjustable square (Figure 10-8) checks squareness and can also measure the amount of variation from true square. First, bring the square and workpiece together; then, using the micrometer dial, adjust the square until the square blade has full contact with the work. The micrometer reading indicates the amount of variation at the top of the blade. The adjustable square is generally used only on a surface plate or similar flat surface.

CYLINDRICAL SQUARES

The cylindrical square (Figure 10-9) is another form of surface plate square commonly found in the toolroom. This square is available in three variations, the plain cylindrical square, the graduated cylindrical square, and the magnetic cylindrical square.

Figure 10-8 Adjustable square. (Courtesy of DoAll Co.)

The plain cylindrical square, as illustrated, is simply a cylindrical tube that is precision ground and lapped and has the base exactly 90° from the sides of the tube. This square is very useful for checking perpendicularity on the surface plate.

The graduated cylindrical square is almost identical to the plain square except that it has a series of dots engraved into the side of the tube. This is read by aligning the square and the work and, by reading the dots, noting the variation from true squareness. Each dot is equal to 0.0002 in.

The magnetic cylindrical square uses a magnetic base to hold the square on the workpiece. The rod is positioned exactly 90° from the base. A dial indicator is then used to check the squareness of the workpiece (Figure 10-10).

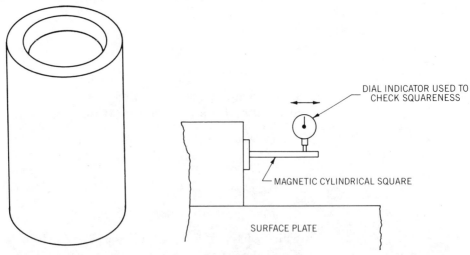

Figure 10-9 Plain cylindrical square.

Figure 10-10 Using the magnetic cylindrical square.

SELF TEST

1. What is the most common level in the shop? (Circle one.)

   a. Electronic type

   b. Bubble type

   c. Lever type

   d. Counterweight type

2. Which of the following levels is the most accurate? (Circle one.)

   a. Pocket level

   b. Machinists' level

   c. Precision level

   d. Cross test level

3. Which of the following is the least accurate level? (Circle one.)

   a. Pocket level

   b. Machinists' level

   c. Precision level

   d. Cross test level

4. Which type of level is best suited for leveling in all directions at one setting? (Circle one.)

   a. Precision level

   b. Machinists' level

   c. Pocket level

   d. Bulls-eye level

5. How accurate is the precision level? (Circle one.)

   a. 0.00005 in. in 12 in.

   b. 0.0005 in. in 12 in.

   c. 0.005 in. in 12 in.

   d. 0.05 in. in 12 in.

6. What type of gage can be used with a straightedge to measure the amount of variation from true flatness? (Circle one.)

   a. Telescoping gage

   b. Feeler gage

   c. Pin gage

   d. Taper gage

7. Which of the following squares can be used for checking both internal and external right angles? (Circle one.)

   a. Solid square

b. Cylindrical square

c. Adjustable square

d. Magnetic cylindrical square

8. Which of the following can measure the variation from true square? (Circle one.)

   a. Plain cylindrical square

   b. Magnetic cylindrical square

   c. Solid square

   d. Adjustable square

9. What is the value of each dot on a graduated cylindrical square? (Circle one.)

   a. 0.00002 in.

   b. 0.0002 in.

   c. 0.002 in.

   d. None of the above

10. Which of the following can measure the angle of a workpiece that is not square? (Circle one.)

    a. Graduated cylindrical square

    b. Magnetic cylindrical square

    c. Adjustable square

    d. None of the above

Answers To Self Test

1. b
2. c
3. a
4. d
5. b
6. b
7. a
8. d
9. b
10. d

SECTION V: ANGULAR MEASURING TOOLS

UNIT 11 Nonprecision Angular Measurements

UNIT 12 Precision Angular Measurements

UNIT 11

Nonprecision Angular Measurements

As a machinist, you will make a countless number of measurements. In addition to the linear measurements already discussed, you will also make many angular measurements. Angular measurement, for the purpose of this book, is divided into two categories, nonprecision angular measurement and precision angular measurement. Nonprecision angular measurement is the measurement of angles to an accuracy of ±1°. Precision angular measurement is the measurement of angles to an accuracy closer than 1°.

WHAT IS ANGULAR MEASUREMENT?

Angles and angular part features are measured in units of degrees, minutes, and seconds. A circle contains 360 degrees, a degree (°) contains 60 minutes, and a minute (′) contains 60 seconds (″) (Figure 11-1).

The circle forms the basis for angular measurement. In fact, all angular measuring tools have circular scales, such as the bevel protractor for the combination square (Figure 11-2). These circular scales generally show 180°, or half a circle. Also notice that the scales on this tool read in both directions. The midpoint on the scale is read the same in both directions, 90°.

As a machinist, you will generally work with angles less than 90°. This is primarily because of the way parts are dimensioned on prints. For example, the parts shown in Figure 11-3 are both dimensioned correctly; however, the

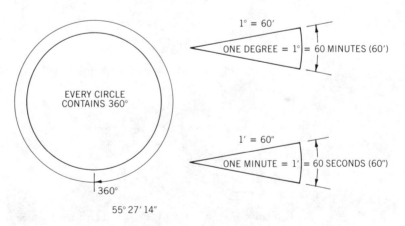

Figure 11-1 Units of angular measurement.

129

130 SECTION V ANGULAR MEASURING TOOLS

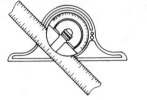

Figure 11-2 Angular measuring tools use circular scales.

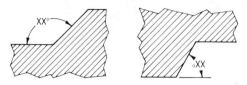

Figure 11-3 Dimensioning angular features.

part on the left would be difficult to measure. For this reason the part is dimensioned as shown on the right. This makes the part easier to measure and lessens the chance of measurement error.

Angular measurements are generally made from either a horizontal or vertical plane (Figure 11-4). When making any angular measurement, you must measure from the correct plane. For example, if a part were dimensioned from the vertical plane and you made your measurement from the horizontal plane, your measurement would be incorrect. Instead of measuring the 50° called for on the print, your measurement would read 40° (Figure 11-5).

Occasionally you will have to make a measurement from the opposite planes due to the shape of the part. In this case you must remember to add or subtract your reading from 90° or 180° to find the correct angle. A few examples of this process are shown in Figure 11-6.

Angular features are sometimes dimensioned in ways that would be impossible to measure directly. In these cases any parallel surface can be used for checking the angle (Figure 11-7). Now let's try a few practice problems to see how well you can calculate angular dimensions (see Figure 11-8).

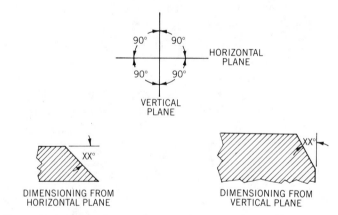

Figure 11-4 Dimensioning from the horizontal or vertical plane.

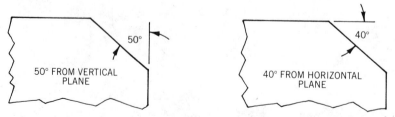

Figure 11-5 Incorrect reading can result from measuring from the wrong plane.

Unit 11 Nonprecision Angular Measurements 131

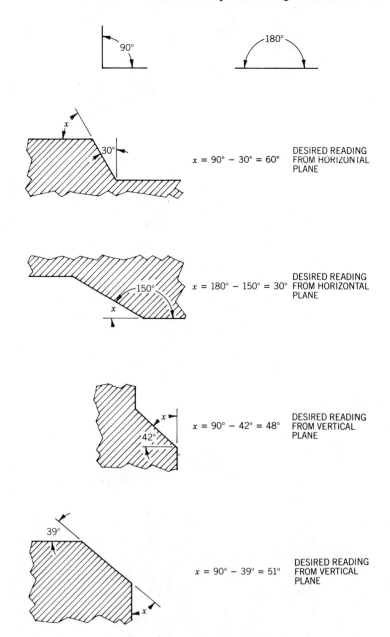

Figure 11-6 Calculating angles.

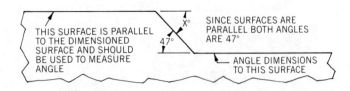

Figure 11-7 Using a parallel surface to make a measurement.

SECTION V ANGULAR MEASURING TOOLS

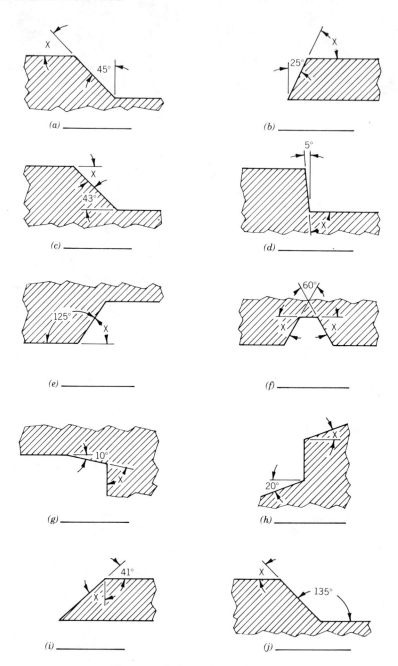

Figure 11-8 Practice problem 1.

PROTRACTORS Protractors are available in several types. In the machine shop the most common types are the bevel protractor, the steel protractor, and the combination protractor and depth gage.

BEVEL PROTRACTORS
The bevel protractor (Figure 11-9) is a versatile part of the combination square. As shown in Figure 11-10, this tool is useful for a variety of measuring applications.

Unit 11 Nonprecision Angular Measurements 133

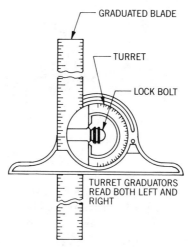

Figure 11-9 Bevel protractor.

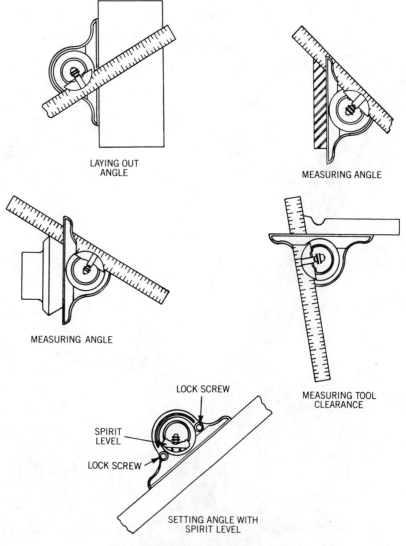

Figure 11-10 Using the bevel protractor.

134 SECTION V ANGULAR MEASURING TOOLS

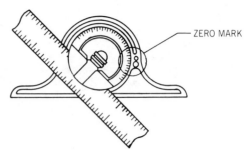

Figure 11-11 Reading a bevel protractor.

When reading a bevel protractor, align the degree marks with the zero mark on the protractor frame (Figure 11-11). The scale you should read depends on whether you want the angle above or below the blade. The measurement is read directly from the circular scale.

STEEL PROTRACTORS

The steel protractor (Figure 11-12) is available in two basic styles, the rectangular head and the semicircular head. These protractors are useful for

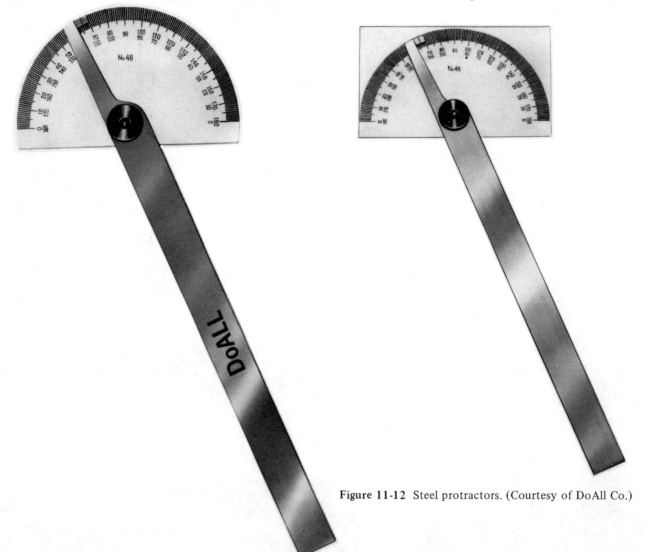

Figure 11-12 Steel protractors. (Courtesy of DoAll Co.)

checking angles in areas where the bevel protractor cannot fit and for angular layout work.

The steel protractor has a fixed scale and a movable blade. The indicating end of the blade has the reference line positioned so angles can be measured on either side of the blade, since the blade cannot be extended or retracted to make measurements in confined places.

COMBINATION PROTRACTOR AND DEPTH GAGE

The combination protractor and depth gage (Figure 11-13) is a simple variation of the steel protractor. Instead of using a fixed blade, this tool has a fixed pivot point and a sliding blade.

The combination protractor and depth gage uses a 6-in. (150-mm) narrow rule as the measuring blade. This permits the tool to make measurements of both angles and depths in areas where the bevel protractor or steel protractor will not fit. Only one side of the blade in this tool can be used for angular

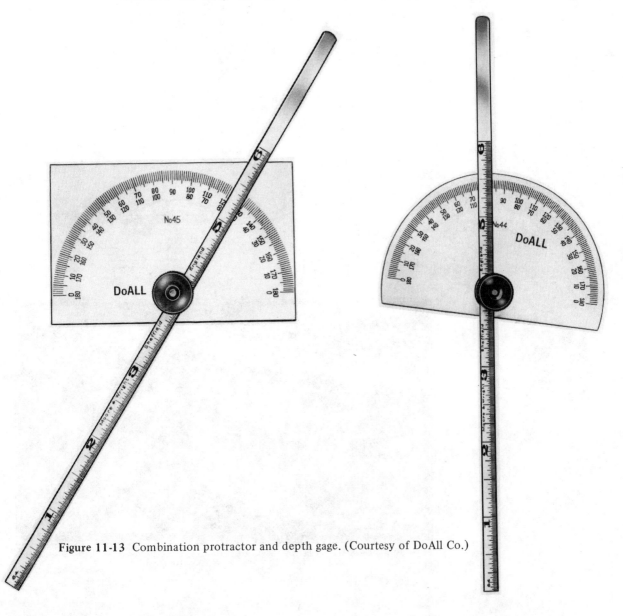

Figure 11-13 Combination protractor and depth gage. (Courtesy of DoAll Co.)

measurements as a result of the off-center design of the pivot screw. To use this protractor, take the readings from the edge of the blade that is aligned with the graduations on the protractor scale.

ANGLE TRANSFER TOOLS

The principal tool used for transfering angular settings is the bevel. Figure 11-14 shows the three principle types of bevel, the plain bevel, the universal bevel, and the combination bevel.

The main purpose of these tools is to check angles in areas where no other tool will work. Once set on the angle being measured, the bevel is removed and checked with a protractor. Figure 11-15 shows several applications where these tools are well suited. Another attachment that greatly extends the usefulness of the bevel is the protractor head (Figure 11-16). This protractor head allows the bevel to make direct, not transfer, measurements.

PLAIN UNIVERSAL COMBINATION

Figure 11-14 Bevels. (Courtesy of L.S. Starrett Co.)

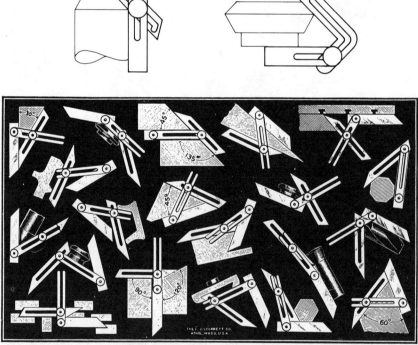

Starrett Universal Bevels are very useful for checking, laying out, and transferring angles on a wide variety of work. Some of the many uses are illustrated above.

Figure 11-15 Typical uses of the bevel. (Courtesy of L.S. Starrett Co.)

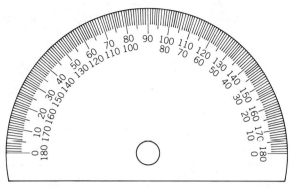

Figure 11-16 Protractor head.

SELF TEST

1. How accurate is nonprecision angular measurement? (Circle one.)
   a. ±0.1°
   b. ±10°
   c. ±0.01°
   d. ±1°

2. How many degrees are there in a circle? (Circle one.)
   a. 360°
   b. 180°
   c. 90°
   d. 45°

3. How many minutes are there in a degree? (Circle one.)
   a. 6'
   b. 60'
   c. 100'
   d. 1000'

4. How many seconds are there in a minute? (Circle one.)
   a. 1000"
   b. 100"
   c. 60"
   d. 6"

5. What type of scale is used for angular measuring tools? (Circle one.)
   a. Round
   b. Rectangular
   c. Circular
   d. Linear

138　SECTION V　ANGULAR MEASURING TOOLS

6. How many degrees are usually found on angular measuring scales? (Circle one.)

    a. 360°

    b. 270°

    c. 180°

    d. 90°

7. How are angular measurements made? (Circle one.)

    a. From the angular surface itself

    b. From a horizontal or vertical plane

    c. From any parallel surface

    d. None of the above

8. List the three types of protractors commonly found in the machine shop.

9. List the three types of bevels commonly found in the machine shop.

10. Find angle "X" for the problems in Figure 11-17.

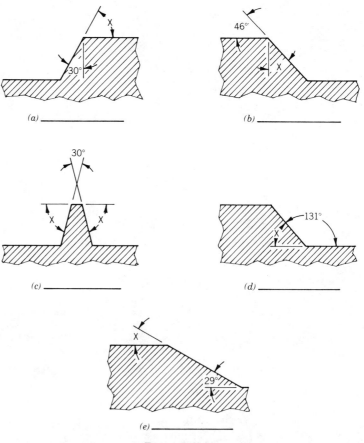

Figure 11-17

Answers to Self Test

1. d
2. a
3. b
4. c
5. c
6. c
7. b
8. a. Bevel protractor
   b. Steel protractor
   c. Combination protractor and depth gage
9. a. Plain bevel
   b. Universal bevel
   c. Combination bevel
10. a. 60°
    b. 44°
    c. 75°
    d. 49°
    e. 29°

Answers to Practice Problem

a. 45°
b. 65°
c. 43°
d. 85°
e. 55°
f. 60°
g. 80°
h. 20°
i. 49°
j. 45°

UNIT 12

Precision Angular Measurements

Precision angular measuring tools are used for measuring angles to accuracies beyond the capabilities of ordinary protractors. While accuracies of ±1° are acceptable for some workpieces, others require accuracy in the range of seconds of a degree. To meet this need, angular measuring tools have been developed that can precisely divide the degree into as many as 14,000 equal parts.

Before discussing these precision angular measuring tools, let's review how angles are divided and how to calculate values in degrees (°), minutes ('), and seconds (").

CALCULATING ANGULAR MEASUREMENTS

Most angles are dimensioned from either a horizontal or vertical plane and each angular surface has two different angles which, for the purpose of this book, will be identified as the working angle and the complementary angle (Figure 12-1). The working angle is the angle that is dimensioned. The complementary angle is the angle between the surface and the opposite plane. In most cases, when making angular measurements, the angle can be measured directly from the working angle. There are times, however, when the complementary angle must be checked to find the working angle. In these cases simply follow the process used in Unit 11, with the following variations.

When calculating precision angular measurements, you will often work with the minutes and seconds of a degree. To calculate these values, you must remember there are 60' in each degree and 60" in each minute. For example, if you wanted to find the complementary angle of the part shown in Figure 12-2, you could take the value 26° 15' and subtract it from 90°. But you will first need to convert 90° to degrees and minutes. 90° = 89° 60'. The complementary angle of this part is 63° 45'.

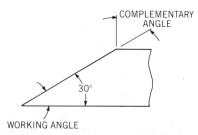

Figure 12-1 Working angles and complementary angles.

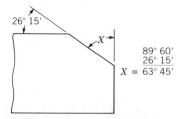

Figure 12-2 Calculating angles in degrees and minutes.

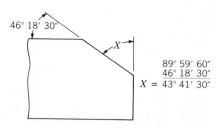

Figure 12-3 Calculating angles in degrees, minutes, and seconds.

Unit 12 Precision Angular Measurements 141

If the working angle were dimensioned as shown in Figure 12-3, the 90° angle would have to be converted to 89° 59′ 60″. The complementary angle in this example is 43° 41′ 30″. Now let's try a few practice problems (see Figure 12-4).

Practice Problem 1

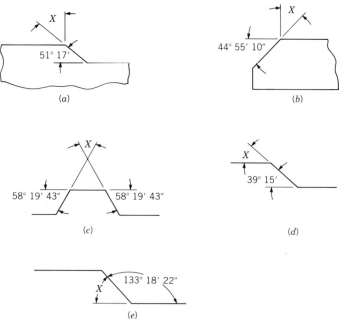

Figure 12-4 Practice problem 1.

THE VERNIER BEVEL PROTRACTOR

The vernier bevel protractor (Figure 12-5) is the most commonly used tool for precision angular measurement. This instrument combines a 360° protractor with a circular vernier scale to divide each degree into 5′ increments.

The 360° dial is further divided into four 90° quadrants that are graduated from 0 to 90° and back to 0° on both sides of the main scale. This permits the scale to be read in either direction. The vernier scale is divided into 24 parts, 12 on each side of the zero. When read together, the main scale and vernier scale have a discrimination of $\frac{1}{12}°$, or 5′.

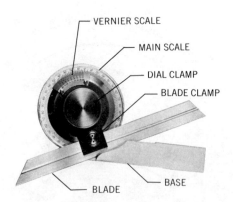

Figure 12-5 Vernier bevel protractor.

142 SECTION V ANGULAR MEASURING TOOLS

READING A VERNIER BEVEL PROTRACTOR

When reading a vernier bevel protractor, the first step is to note the location and direction of the zero on the vernier scale. The direction of movement of the dial is important because this determines the position of the reading on the vernier scale. As a rule, you should always read the vernier in the same direction from zero as you read the main dial (Figure 12-6).

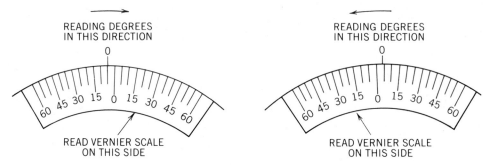

Figure 12-6 Reading the direction of travel.

When reading the vernier bevel protractor, first read the degree graduations either to the left or right of the zero on the vernier scale, depending on the direction of the measurement. Next, find the line on the vernier scale that is aligned with a line on the main scale and add this value to the degree reading (Figure 12-7). Always remember to read the correct side of the vernier scale.

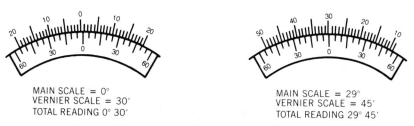

Figure 12-7 Reading the vernier bevel protractor.

Now let's try a few practice problems in reading the vernier bevel protractor (see Figure 12-8).

Practice Problem 2

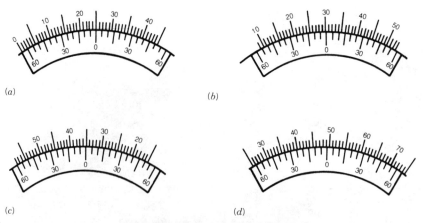

Figure 12-8 Practice problem 2.

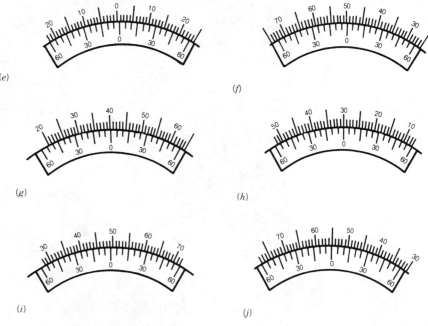

Figure 12-8 (continued)

USING A VERNIER BEVEL PROTRACTOR

The vernier bevel protractor is a useful and versatile tool but, like all precision measuring tools, it must be used properly to give the desired results. The following are a few points to keep in mind when using vernier bevel protractors.

1. Make sure the workpiece is clean and free of burrs.
2. Make sure the protractor is clean and in good condition; examine it for signs of damage prior to use.

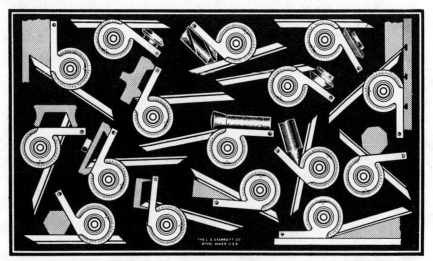

Starrett Universal Bevel Protractors are invaluable tools for precision measuring and laying out of angles. Typical applications are shown above.

Figure 12-9 Typical applications of the vernier bevel protractor. (Courtesy of L.S. Starrett Co.)

144 SECTION V ANGULAR MEASURING TOOLS

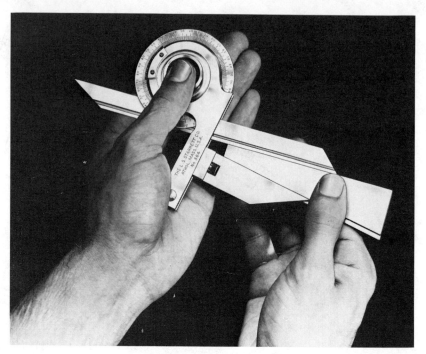

Figure 12-10 Using a vernier bevel protractor with an acute angle attachment. (Courtesy of L.S. Starrett Co.)

3. Do not slide the protractor along an abrasive surface.
4. Do not overtighten the clamp screws.
5. Make sure the protractor and workpiece are properly aligned.
6. Make sure the protractor blade makes complete contact with the workpiece.
7. Keep the protractor in its case when not in use.
8. Wipe off the protractor and coat it with a thin film of corrosion-resistant compound before returning it to the toolroom.

Figure 12-9 shows several examples of how vernier bevel protractors are used. In cases where a part cannot be measured with the protractor alone, use an acute angle attachment (Figure 12-10).

THE SINE BAR

The sine bar (Figure 12-11) is a precision-made tool used for a variety of measuring and machining applications. The basic construction of the sine bar consists of a rectangular, or square, bar that is fitted with a cylindrical plug at each end. The bar and the cylindrical plugs are precision ground and lapped to an accuracy of 0.0001 in. in length and 0.000025 in. in parallel. The sine bar is a precision instrument and should always be treated carefully.

Figure 12-11 Sine bar.

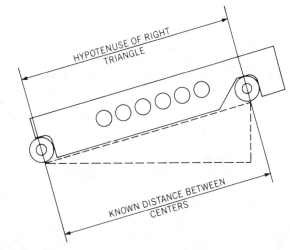

Figure 12-12 The sine bar replaces the hypotenuse in a right triangle.

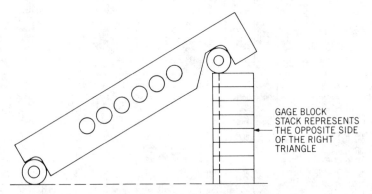

Figure 12-13 The gage block stack replaces the opposite side in a right triangle.

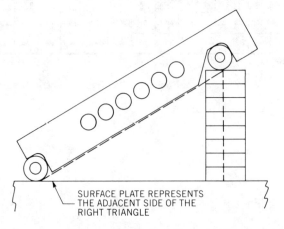

Figure 12-14 A surface plate represents the adjacent side of a right triangle.

The sine bar is used to replace the hypotenuse, or longest side, of a right triangle (Figure 12-12). The cylindrical plugs are positioned at an exact, known distance apart, generally 5 or 10 in. for the inch size bars or 100 or 200 mm for the metric sine bars. The opposite side of the right triangle is determined by a gage block stack (Figure 12-13). A surface plate is used to represent the adjacent side of the triangle (Figure 12-14).

NATURAL TRIGONOMETRIC FUNCTIONS *continued*

	20°						21°						22°						23°						
′	sin	cos	tan	cot	sec	cosec	sin	cos	tan	cot	sec	cosec	sin	cos	tan	cot	sec	cosec	sin	cos	tan	cot	sec	cosec	′
0	.34202	.93969	.36397	2.7475	1.0642	2.9238	.35837	.93358	.38386	2.6051	1.0711	2.7904	.37461	.92718	.40403	2.4751	1.0785	2.6695	.39073	.92050	.42447	2.3558	1.0864	2.5593	60
1	.34229	.93959	.36430	.7450	.0643	.9215	.35864	.93348	.38420	.6028	.0713	.7883	.37488	.92707	.40436	.4730	.0787	.6675	.39100	.92039	.42482	.3539	.0865	.5575	59
2	.34257	.93949	.36463	.7425	.0644	.9191	.35891	.93337	.38453	.6006	.0714	.7862	.37514	.92696	.40470	.4709	.0788	.6656	.39126	.92028	.42516	.3520	.0866	.5558	58
3	.34284	.93939	.36496	.7400	.0645	.9168	.35918	.93327	.38486	.5983	.0715	.7841	.37541	.92686	.40504	.4689	.0789	.6637	.39153	.92016	.42550	.3501	.0868	.5540	57
4	.34311	.93929	.36529	.7376	.0646	.9145	.35945	.93316	.38520	.5960	.0716	.7820	.37568	.92675	.40538	.4668	.0790	.6618	.39180	.92005	.42585	.3482	.0869	.5523	56
5	.34339	.93919	.36562	2.7351	.0647	2.9122	.35972	.93306	.38553	2.5938	.0717	2.7799	.37595	.92664	.40572	2.4647	.0792	2.6599	.39207	.91993	.42619	2.3463	.0870	2.5506	55
6	.34366	.93909	.36595	.7326	.0648	.9098	.36000	.93295	.38587	.5916	.0719	.7778	.37622	.92653	.40606	.4627	.0793	.6580	.39234	.91982	.42654	.3445	.0872	.5488	54
7	.34393	.93899	.36628	.7302	.0650	.9075	.36027	.93285	.38620	.5893	.0720	.7757	.37649	.92642	.40640	.4606	.0794	.6561	.39260	.91971	.42688	.3426	.0873	.5471	53
8	.34421	.93889	.36661	.7277	.0651	.9052	.36054	.93274	.38654	.5871	.0721	.7736	.37676	.92631	.40673	.4586	.0795	.6542	.39287	.91959	.42722	.3407	.0874	.5453	52
9	.34448	.93879	.36694	.7252	.0652	.9029	.36081	.93264	.38687	.5848	.0722	.7715	.37703	.92620	.40707	.4565	.0797	.6523	.39314	.91948	.42757	.3388	.0876	.5436	51
10	.34475	.93869	.36727	2.7228	.0653	2.9006	.36108	.93253	.38720	2.5826	.0723	2.7694	.37730	.92609	.40741	2.4545	.0798	2.6504	.39341	.91936	.42791	2.3369	.0877	2.5419	50
11	.34502	.93859	36760	.7204	.0654	.8983	.36135	.93243	.38754	.5804	.0725	.7674	.37757	.92598	.40775	.4525	.0799	.6485	.39367	.91925	.42826	.3350	.0878	.5402	49
12	.34530	.93849	.36793	.7179	.0655	.8960	.36162	.93232	.38787	.5781	.0726	.7653	.37784	.92587	.40809	.4504	.0801	.6466	.39394	.91913	.42860	.3332	.0880	.5384	48
13	.34557	.93839	.36826	.7155	.0656	.8937	.36189	.93222	.38821	.5759	.0727	.7632	.37811	.92576	.40843	.4484	.0802	.6447	.39421	.91902	.42894	.3313	.0881	.5367	47
14	.34584	.93829	.36859	.7130	.0658	.8915	.36217	.93211	.38854	.5737	.0728	.7611	.37838	.92565	.40877	.4463	.0803	.6428	.39448	.91891	.42929	.3294	.0882	.5350	46
15	.34612	.93819	.36892	2.7106	.0659	2.8892	.36244	.93201	.38888	2.5715	.0729	2.7591	.37865	.92554	.40911	2.4443	.0804	2.6410	.39474	.91879	.42963	2.3276	.0884	2.5333	45
16	.34639	.93809	.36925	.7082	.0660	.8869	.36271	.93190	.38921	.5693	.0731	.7570	.37892	.92543	.40945	.4423	.0806	.6391	.39501	.91868	.42998	.3257	.0885	.5316	44
17	.34666	.93799	.36958	.7058	.0661	.8846	.36298	.93180	.38955	.5671	.0732	.7550	.37919	.92532	.40979	.4403	.0807	.6372	.39528	.91856	.43032	.3238	.0886	.5299	43
18	.34693	.93789	.36991	.7033	.0662	.8824	.36325	.93169	.38988	.5649	.0733	.7529	.37946	.92521	.41013	.4382	.0808	.6353	.39554	.91845	.43067	.3220	.0888	.5281	42
19	.34721	.93779	.37024	.7009	.0663	.8801	.36352	.93158	.39022	.5627	.0734	.7509	.37972	.92510	.41047	.4362	.0810	.6335	.39581	.91833	.43101	.3201	.0889	.5264	41
20	.34748	.93769	.37057	2.6985	.0664	2.8778	.36379	.93148	.39055	2.5605	.0736	2.7488	.37999	.92499	.41081	2.4342	.0811	2.6316	.39608	.91822	.43136	2.3183	.0891	2.5247	40
21	.34775	.93758	.37090	.6961	.0666	.8756	.36406	.93137	.39089	.5583	.0737	.7468	.38026	.92488	.41115	.4322	.0812	.6297	.39635	.91810	.43170	.3164	.0892	.5230	39
22	.34803	.93748	.37123	.6937	.0667	.8733	.36433	.93127	.39122	.5561	.0738	.7447	.38053	.92477	.41149	.4302	.0813	.6279	.39661	.91798	.43205	.3145	.0893	.5213	38
23	.34830	.93738	.37156	.6913	.0668	.8711	.36460	.93116	.39156	.5539	.0739	.7427	.38080	.92466	.41183	.4282	.0815	.6260	.39688	.91787	.43239	.3127	.0895	.5196	37
24	.34857	.93728	.37190	.6889	.0669	.8688	.36488	.93105	.39189	.5517	.0740	.7406	.38107	.92455	.41217	.4262	.0816	.6242	.39715	.91775	.43274	.3109	.0896	.5179	36
25	.34884	.93718	.37223	2.6865	.0670	2.8666	.36515	.93095	.39223	2.5495	.0742	2.7386	.38134	.92443	.41251	2.4242	.0817	2.6223	.39741	.91764	.43308	2.3090	.0897	2.5163	35
26	.34912	.93708	.37256	.6841	.0671	.8644	.36542	.93084	.39257	.5473	.0743	.7366	.38161	.92432	.41285	.4222	.0819	.6205	.39768	.91752	.43343	.3072	.0899	.5146	34
27	.34939	.93698	.37289	.6817	.0673	.8621	.36569	.93074	.39290	.5451	.0744	.7346	.38188	.92421	.41319	.4202	.0820	.6186	.39795	.91741	.43377	.3053	.0900	.5129	33
28	.34966	.93687	.37322	.6794	.0674	.8599	.36596	.93063	.39324	.5430	.0745	.7325	.38214	.92410	.41353	.4182	.0821	.6168	.39821	.91729	.43412	.3035	.0902	.5112	32
29	.34993	.93677	.37355	.6770	.0675	.8577	.36623	.93052	.39357	.5408	.0747	.7305	.38241	.92399	.41387	.4162	.0823	.6150	.39848	.91718	.43447	.3017	.0903	.5095	31
30	.35021	.93667	.37388	2.6746	.0676	2.8554	.36650	.93042	.39391	2.5386	.0748	2.7285	.38268	.92388	.41421	2.4142	.0824	2.6131	.39875	.91706	.43481	2.2998	.0904	2.5078	30
31	.35048	.93657	.37422	.6722	.0677	.8532	.36677	.93031	.39425	.5365	.0749	.7265	.38295	.92377	.41455	.4122	.0825	.6113	.39901	.91694	.43516	.2980	.0906	.5062	29
32	.35075	.93647	.37455	.6699	.0678	.8510	.36704	.93020	.39458	.5343	.0750	.7245	.38322	.92366	.41489	.4102	.0826	.6095	.39928	.91683	.43550	.2962	.0907	.5045	28
33	.35102	.93637	.37488	.6675	.0679	.8488	.36731	.93010	.39492	.5322	.0751	.7225	.38349	.92354	.41524	.4083	.0828	.6076	.39955	.91671	.43585	.2944	.0908	.5028	27
34	.35130	.93626	.37521	.6652	.0681	.8466	.36758	.92999	.39525	.5300	.0753	.7205	.38376	.92343	.41558	.4063	.0829	.6058	.39981	.91659	.43620	.2925	.0910	.5011	26
35	.35157	.93616	.37554	2.6628	.0682	2.8444	.36785	.92988	.39559	2.5278	.0754	2.7185	.38403	.92332	.41592	2.4043	.0830	2.6040	.40008	.91648	.43654	2.2907	.0911	2.4995	25
36	.35184	.93606	.37587	.6604	.0683	.8422	.36812	.92978	.39593	.5257	.0755	.7165	.38429	.92321	.41626	.4023	.0832	.6022	.40035	.91636	.43689	.2889	.0913	.4978	24
37	.35211	.93596	.37621	.6581	.0684	.8400	.36839	.92967	.39626	.5236	.0756	.7145	.38456	.92310	.41660	.4004	.0833	.6003	.40061	.91625	.43723	.2871	.0914	.4961	23
38	.35239	.93585	.37654	.6558	.0685	.8378	.36866	.92956	.39660	.5214	.0758	.7125	.38483	.92299	.41694	.3984	.0834	.5985	.40088	.91613	.43758	.2853	.0915	.4945	22
39	.35266	.93575	.37687	.6534	.0686	.8356	.36893	.92945	.39694	.5193	.0759	.7105	.38510	.92287	.41728	.3964	.0836	.5967	.40115	.91601	.43793	.2835	.0917	.4928	21
40	.35293	.93565	.37720	2.6511	.0688	2.8334	.36921	.92935	.39727	2.5171	.0760	2.7085	.38537	.92276	.41762	2.3945	.0837	2.5949	.40141	.91590	.43827	2.2817	.0918	2.4912	20
41	.35320	.93555	.37754	.6487	.0689	.8312	.36948	.92924	.39761	.5150	.0761	.7065	.38564	.92265	.41797	.3925	.0838	.5931	.40168	.91578	.43862	.2799	.0920	.4895	19
42	.35347	.93544	.37787	.6464	.0690	.8290	.36975	.92913	.39795	.5129	.0763	.7045	.38591	.92254	.41831	.3906	.0840	.5913	.40195	.91566	.43897	.2781	.0921	.4879	18
43	.35375	.93534	.37820	.6441	.0691	.8269	.37002	.92902	.39828	.5108	.0764	.7026	.38617	.92242	.41865	.3886	.0841	.5895	.40221	.91554	.43932	.2763	.0922	.4862	17
44	.35402	.93524	.37853	.6418	.0692	.8247	.37029	.92892	.39862	.5086	.0765	.7006	.38644	.92231	.41899	.3867	.0842	.5877	.40248	.91543	.43966	.2745	.0924	.4846	16
45	.35429	.93513	.37887	2.6394	.0694	2.8225	.37056	.92881	.39896	2.5065	.0766	2.6986	.38671	.92220	.41933	2.3847	.0844	2.5958	.40205	.91531	.44001	2.2727	.0925	2.4829	15
46	.35456	.93503	.37920	.6371	.0695	.8204	.37083	.92870	.39930	.5044	.0768	.6967	.38698	.92209	.41968	.3828	.0845	.5841	.40301	.91519	.44036	.2709	.0927	.4813	14
47	.35483	.93493	.37953	.6348	.0696	.8182	.37110	.92859	.39963	.5023	.0769	.6947	.38725	.92197	.42002	.3808	.0846	.5823	.40328	.91508	.44070	.2691	.0928	.4797	13
48	.35511	.93482	.37986	.6325	.0697	.8160	.37137	.92848	.39997	.5002	.0770	.6927	.38751	.92186	.42036	.3789	.0847	.5805	.40354	.91496	.44105	.2673	.0929	.4780	12
49	.35538	.93472	.38020	.6302	.0698	.8139	.37164	.92838	.40031	.4981	.0771	.6908	.38778	.92105	.42070	.3770	.0849	.5787	.40381	.91484	.44140	.2655	.0931	.4764	11
50	.35565	.93462	.38053	2.6279	.0699	2.8117	.37191	.92827	.40065	2.4960	.0773	2.6888	.38805	.92164	.42105	2.3750	.0850	2.5770	.40408	.91472	.44175	2.2637	.0932	2.4748	10
51	.35592	.93451	.38086	.6256	.0701	.8096	.37218	.92816	.40098	.4939	.0774	.6869	.38832	.92152	.42139	.3731	.0851	.5752	.40434	.91461	.44209	.2619	.0934	.4731	9
52	.35619	.93441	.38120	.6233	.0702	.8074	.37245	.92805	.40132	.4918	.0775	.6849	.38859	.92141	.42173	.3712	.0853	.5734	.40461	.91449	.44244	.2602	.0935	.4715	8
53	.35647	.93431	.38153	.6210	.0703	.8053	.37272	.92794	.40166	.4897	.0776	.6830	.38886	.92130	.42207	.3692	.0854	.5716	.40487	.91437	.44279	.2584	.0936	.4699	7
54	.35674	.93420	.38186	.6187	.0704	.8032	.37299	.92784	.40200	.4876	.0778	.6810	.38912	.92118	.42242	.3673	.0855	.5699	.40514	.91425	.44314	.2566	.0938	.4683	6
55	.35701	.93410	.38220	2.6164	.0705	2.8010	.37326	.92773	.40233	2.4855	.0779	2.6791	.38939	.92107	.42276	2.3654	.0857	2.5681	.40541	.91414	.44349	2.2548	.0939	2.4666	5
56	.35728	.93400	.38253	.6142	.0707	.7989	.37353	.92762	.40267	.4834	.0780	.6772	.38966	.92096	.42310	.3635	.0858	.5663	.40567	.91402	.44383	.2531	.0941	.4650	4
57	.35755	.93389	.38286	.6119	.0708	.7968	.37380	.92751	.40301	.4813	.0781	.6752	.38993	.92084	.42344	.3616	.0859	.5646	.40594	.91390	.44418	.2513	.0942	.4634	3
58	.35782	.93379	.38320	.6096	.0709	.7947	.37407	.92740	.40335	.4792	.0783	.6733	.39019	.92073	.42379	.3597	.0861	.5628	.40620	.91378	.44453	.2495	.0943	.4618	2
59	.35810	.93368	.38353	.6073	.0710	.7925	.37434	.92729	.40369	.4772	.0784	.6714	.39046	.92062	.42413	.3577	.0862	.5610	.40647	.91366	.44488	.2478	.0945	.4602	1
60	.35837	.93358	.38386	2.6051	.0711	2.7904	.37461	.92718	.40403	2.4751	.0785	2.6695	.39073	.92050	.42447	2.3558	.0864	2.5593	.40674	.91354	.44523	2.2460	.0946	2.4586	0
′	cos	sin	cot	tan	cosec	sec	cos	sin	cot	tan	cosec	sec	cos	sin	cot	tan	cosec	sec	cos	sin	cot	tan	cosec	sec	′
	69°						68°						67°						66°						

Figure 12-15 Trigonometric functions table.

NATURAL TRIGONOMETRIC FUNCTIONS *continued*

	24°						25°						26°						27°						
′	sin	cos	tan	cot	sec	cosec	sin	cos	tan	cot	sec	cosec	sin	cos	tan	cot	sec	cosec	sin	cos	tan	cot	sec	cosec	′
0	.40674	.91354	.44523	2.2460	1.0946	2.4586	.42262	.90631	.46631	2.1445	1.1034	2.3662	.43837	.89879	.48773	2.0503	1.1126	2.2812	.45399	.89101	.50952	1.9626	1.1223	2.2027	60
1	.40700	.91343	.44558	.2443	.0948	.4570	.42288	.90618	.46666	.1429	.1035	.3647	.43863	.89867	.48809	.0488	.1127	.2798	.45425	.89087	.50989	.9612	.1225	.2014	59
2	.40727	.91331	.44593	.2425	.0949	.4554	.42314	.90606	.46702	.1412	.1037	.3632	.43889	.89854	.48845	.0473	.1129	.2784	.45451	.89074	.51026	.9598	.1226	.2002	58
3	.40753	.91319	.44627	.2408	.0951	.4538	.42341	.90594	.46737	.1396	.1038	.3618	.43915	.89841	.48881	.0458	.1131	.2771	.45477	.89061	.51062	.9584	.1228	.1989	57
4	.40780	.91307	.44662	.2390	.0952	.4522	.42367	.90581	.46772	.1380	.1040	.3603	.43942	.89828	.48917	.0443	.1132	.2757	.45503	.89048	.51099	.9570	.1230	.1977	56
5	.40806	.91295	.44697	2.2373	.0953	2.4506	.42394	.90569	.46808	2.1364	.1041	2.3588	.43968	.89815	.48953	2.0427	.1134	2.2744	.45528	.89034	.51136	1.9556	.1231	2.1964	55
6	.40833	.91283	.44732	.2355	.0955	.4490	.42420	.90557	.46843	.1348	.1043	.3574	.43994	.89803	.48989	.0412	.1135	.2730	.45554	.89021	.51172	.9542	.1233	.1952	54
7	.40860	.91271	.44767	.2338	.0956	.4474	.42446	.90544	.46879	.1331	.1044	.3559	.44020	.89790	.49025	.0397	.1137	.2717	.45580	.89008	.51209	.9528	.1235	.1939	53
8	.40886	.91260	.44802	.2320	.0958	.4458	.42473	.90532	.46914	.1315	.1046	.3544	.44046	.89777	.49062	.0382	.1139	.2703	.45606	.88995	.51246	.9514	.1237	.1927	52
9	.40913	.91248	.44837	.2303	.0959	.4442	.42499	.90520	.46950	.1299	.1047	.3530	.44072	.89764	.49098	.0367	.1140	.2690	.45632	.88981	.51283	.9500	.1238	.1914	51
10	.40939	.91236	.44872	2.2286	.0961	2.4426	.42525	.90507	.46985	2.1283	.1049	2.3515	.44098	.89751	.49134	2.0352	.1142	2.2676	.45658	.88968	.51319	1.9486	.1240	2.1902	50
11	.40966	.91224	.44907	.2268	.0962	.4418	.42552	.90495	.47021	.1267	.1050	.3501	.44124	.89739	.49170	.0338	.1143	.2663	.45684	.88955	.51356	.9472	.1242	.1889	49
12	.40992	.91212	.44942	.2251	.0963	.4395	.42578	.90483	.47050	.1251	.1052	.3486	.44150	.89726	.49206	.0323	.1145	.2650	.45710	.88942	.51393	.9458	.1243	.1877	48
13	.41019	.91200	.44977	.2234	.0965	.4379	.42604	.90470	.47092	.1235	.1053	.3472	.44177	.89713	.49242	.0308	.1147	.2636	.45736	.88928	.51430	.9444	.1245	.1865	47
14	.41045	.91188	.45012	.2216	.0966	.4363	.42630	.90458	.47127	.1219	.1055	.3457	.44203	.89700	.49278	.0293	.1148	.2623	.45761	.88915	.51466	.9430	.1247	.1852	46
15	.41072	.91176	.45047	2.2199	.0968	2.4347	.42657	.90445	.47163	2.1203	.1056	2.3443	.44229	.89687	.49314	2.0278	.1150	2.2610	.45787	.88902	.51503	1.9416	.1248	2.1840	45
16	.41098	.91164	.45082	.2182	.0969	.4332	.42683	.90433	.47199	.1187	.1058	.3428	.44255	.89674	.49351	.0263	.1151	.2596	.45813	.88888	.51540	.9402	.1250	.1828	44
17	.41125	.91152	.45117	.2165	.0971	.4316	.42709	.90421	.47234	.1171	.1059	.3414	.44281	.89661	.49387	.0248	.1153	.2583	.45839	.88875	.51577	.9388	.1252	.1815	43
18	.41151	.91140	.45152	.2147	.0972	.4300	.42736	.90408	.47270	.1155	.1061	.3399	.44307	.89649	.49423	.0233	.1155	.2570	.45865	.88862	.51614	.9375	.1253	.1803	42
19	.41178	.91128	.45187	.2130	.0973	.4285	.42762	.90396	.47305	.1139	.1062	.3385	.44333	.89636	.49459	.0219	.1156	.2556	.45891	.88848	.51651	.9361	.1255	.1791	41
20	.41204	.91116	.45222	2.2113	.0975	2.4269	.42788	.90383	.47341	2.1123	.1064	2.3371	.44359	.89623	.49495	2.0204	.1158	2.2543	.45917	.88835	.51687	1.9347	.1257	2.1778	40
21	.41231	.91104	.45257	.2096	.0976	.4254	.42815	.90371	.47376	.1107	.1065	.3356	.44385	.89610	.49532	.0189	.1159	.2530	.45942	.88822	.51724	.9333	.1258	.1766	39
22	.41257	.91092	.45292	.2079	.0978	.4238	.42841	.90358	.47412	.1092	.1067	.3342	.44411	.89597	.49568	.0174	.1161	.2517	.45968	.88808	.51761	.9319	.1260	.1754	38
23	.41284	.91080	.45327	. 2062	.0979	.4222	.42867	.90346	.47448	.1076	.1068	.3328	.44437	.89584	.49604	.0159	.1163	.2503	.45994	.88795	.51798	.9306	.1262	.1742	37
24	.41310	.91068	.45362	.2045	.0981	.4207	.42893	.90333	.47483	.1060	.1070	.3313	.44463	.89571	.49640	.0145	.1164	.2490	.46020	.88781	.51835	.9292	.1264	.1730	36
25	.41337	.91056	.45397	2.2028	.0982	2.4191	42920	.90321	.47519	2.1044	.1072	2.3299	.44489	.89558	.49677	2.0130	.1166	2.2477	.46046	.88768	.51872	1.9278	.1265	2.1717	35
26	.41363	.91044	.45432	.2011	.0984	.4176	.42946	.90308	.47555	.1028	.1073	.3285	.44516	.89545	.49713	.0115	.1167	.2464	.46072	.88755	.51909	.9264	.1267	.1705	34
27	.41390	.91032	.45467	.1994	.0985	.4160	.42972	.90296	.47590	.1013	.1075	.3271	.44542	.89532	.49749	.0101	.1169	.2451	.46097	.88741	.51946	.9251	.1269	.1693	33
28	.41416	.91020	.45502	.1977	.0986	.4145	.42998	.90283	.47626	.0997	.1076	.3256	.44568	.89519	.49785	.0086	.1171	.2438	.46123	.88728	.51983	.9237	.1270	.1681	32
29	.41443	.91008	.45537	.1960	.0988	.4130	.43025	.90271	.47662	.0981	.1078	.3242	.44594	.89506	.49822	.0071	.1172	.2425	.46149	.88714	.52020	.9223	.1272	.1669	31
30	.41469	.90996	.45573	2.1943	.0989	2.4114	.43051	.90258	.47697	2.0965	.1079	2.3228	.44620	.89493	.49858	2.0057	.1174	2.2411	.46175	.88701	.52057	1.9210	.1274	2.1657	30
31	.41496	.90984	.45608	.1926	.0991	.4099	.43077	.90246	.47733	.0950	.1081	.3214	.44646	.89480	.49894	.0042	.1176	.2398	.46201	.88688	.52094	.9196	.1275	.1645	29
32	.41522	.90972	.45643	.1909	.0992	.4083	.43104	.90233	.47769	.0934	.1082	.3200	.44672	.89467	.49931	.0028	.1177	.2385	.46226	.88674	.52131	.9182	.1277	.1633	28
33	.41549	.90960	.45678	.1892	.0994	.4068	.43130	.90221	.47805	.0918	.1084	.3186	.44698	.89454	.49967	.0013	.1179	.2372	.46252	.88661	.52168	.9169	.1279	.1620	27
34	.41575	.90948	.45713	.1875	.0995	.4053	.43156	.90208	.47840	.0903	.1085	.3172	.44724	.89441	.50003	1.9998	.1180	.2359	.46278	.88647	.52205	.9155	.1281	.1608	26
35	.41602	.90936	.45748	2.1859	.0997	2.4037	.43182	.90196	.47876	2.0887	.1087	2.3158	.44750	.89428	.50040	1.9984	.1182	2.2346	.46304	.88634	.52242	1.9142	.1282	2.1596	25
36	.41628	.90924	.45783	.1842	.0998	.4022	.43208	.90183	.47912	.0872	.1088	.3143	.44776	.89415	.50076	.9969	.1184	.2333	.46330	.88620	.52279	.9128	.1284	.1584	24
37	.41654	.90911	.45819	.1825	.1000	.4007	.43235	.90171	.47948	.0856	.1090	.3129	.44802	.89402	.50113	.9955	.1185	.2320	.46355	.88607	.52316	.9115	.1286	.1572	23
38	.41681	.90899	.45854	.1808	.1001	.3992	.43261	.90158	.47983	.0840	.1092	.3115	.44828	.89389	.50149	.9940	.1187	.2307	.46381	.88593	.52353	.9101	.1287	.1560	22
39	.41707	.90887	.45889	.1792	.1003	.3976	.43287	.90145	.48019	.0825	.1093	.3101	.44854	.89376	.50185	.9926	.1189	.2294	.46407	.88580	.52390	.9088	.1289	.1548	21
40	.41734	.90875	.45924	2.1775	.1004	2.3961	.43313	.90133	.48055	2.0809	.1095	2.3087	.44880	.89363	.50222	1.9912	.1190	2.2282	.46433	.88566	.52427	1.9074	.1291	2.1536	20
41	.41760	.90863	.45960	.1758	.1005	.3946	.43340	.90120	.48091	.0794	.1096	.3073	.44906	.89350	.50258	.9897	.1192	.2269	.46458	.88553	.52464	.9061	.1293	.1525	19
42	.41787	.90851	.45995	.1741	.1007	.3931	.43366	.90108	.48127	.0778	.1098	.3059	.44932	.89337	.50295	.9883	.1193	.2256	.46484	.88539	.52501	.9047	.1294	.1513	18
43	.41813	.90839	.46030	.1725	.1008	.3916	.43392	.90095	.48162	.0763	.1099	.3046	.44958	.89324	.50331	.9868	.1195	.2243	.46510	.88526	.62538	.9034	.1296	.1501	17
44	.41839	.90826	.46065	.1708	.1010	.3901	.43418	.90082	.48198	.0747	.1101	.3032	.44984	.89311	.50368	.9854	.1197	.2230	.46536	.88512	.52575	.9020	.1298	.1489	16
45	.41866	.90814	.46101	2.1692	.1011	2.3886	.43444	.90070	.48234	2.0732	.1102	2.3018	.45010	.89298	.50404	1.9840	.1198	2.2217	.46561	.88499	.52612	1.9007	.1299	2.1477	15
46	.41892	.90812	.46136	.1675	.1013	.3871	.43471	.90057	.48270	.0717	.1104	.3004	.45036	.89285	.50441	.9825	.1200	.2204	.46587	.88485	.52650	.8993	.1301	.1465	14
47	.41919	.90790	.46171	.1658	.1014	.3856	.43497	.90044	.48306	.0701	.1106	.2990	.45062	.89272	.50477	.9811	.1202	.2192	.46613	.88472	.52687	.8980	.1303	.1453	13
48	.41945	.90778	.46206	.1642	.1016	.3841	.43523	.90032	.48342	.0686	.1107	.2976	.45088	.89258	.50514	.9797	.1203	.2179	.46639	.88458	.52724	.8967	.1305	.1441	12
49	.41972	.90765	.46242	.1625	.1017	.3826	.43549	.90019	.48378	.0671	.1109	.2962	.45114	.89245	.50550	.9782	.1205	.2166	.46664	.88444	.52761	.8953	.1306	.1430	11
50	.41998	.90753	.46277	2.1609	.1019	2.3811	.43575	.90006	.48414	2.0655	.1110	2.2949	.45140	.89232	.50587	1.9768	.1207	2.2153	.46690	.88431	.52798	1.8940	.1308	2.1418	10
51	.42024	.90741	.46312	.1592	.1020	.3796	.43602	.89994	.48449	.0640	.1112	.2935	.45166	.89219	.50623	.9754	.1208	.2141	.46716	.88417	.52836	.8927	.1310	.1406	9
52	.42051	.90729	.46348	.1576	.1022	.3781	.43628	.89981	.48485	.0625	.1113	.2921	.45191	.89206	.50660	.9739	.1210	.2128	.46741	.88404	.52873	.8913	.1312	.1394	8
53	.42077	.90717	.46383	.1559	.1023	.3766	.43654	.89968	.48521	.0609	.1115	.2907	.45217	.89193	.50696	.9725	.1212	.2115	.46767	.88390	.52910	.8900	.1313	.1382	7
54	.42103	.90704	.46418	.1543	.1025	.3751	.43680	.89956	.48557	.0594	.1116	.2894	.45243	.89180	.50733	.9711	.1213	.2103	.46793	.88376	.52947	.8887	.1315	.1371	6
55	.42130	.90692	.46454	2.1527	.1026	2.3736	.43706	.89943	.48593	2.0579	.1118	2.2880	.45269	.89166	.50769	1.9697	.1215	2.2090	.46819	.88363	.52984	1.8873	.1317	2.1359	5
56	.42156	.90680	.46489	.1510	.1028	.3721	.43732	.89930	.48629	.0564	.1120	.2866	.45295	.89153	.50806	.9683	.1217	.2077	.46844	.88349	.53022	.8860	.1319	.1347	4
57	.42183	.90668	.46524	.1494	.1029	.3706	.43759	.89918	.48665	.0548	.1121	.2853	.45321	.89140	.50843	.9668	.1218	.2065	.46870	.88336	.53059	..8847	.1320	.1335	3
58	.42209	.90655	.46560	.1478	.1031	.3691	.43785	.89905	.48701	.0533	.1123	.2839	.45347	.89127	.50879	.9654	.1220	.2052	.46896	.88322	.53096	.8834	.1322	.1324	2
59	.42235	.90643	.46595	.1461	.1032	.3677	.43811	.89892	.48737	.0518	.1124	.2825	.45373	.89114	.50916	.9640	.1222	.2039	.46921	.88308	.53134	.8820	.1324	.1312	1
60	.42262	.90631	.46631	2.1445	.1034	2.3662	.43837	.89879	.48773	2.0503	.1126	2.2812	.45399	.89101	.50952	1.9626	.1223	2.2027	.46947	.88295	.53171	1.8807	.1326	2.1300	0
′	cos	sin	cot	tan	cosec	sec	cos	sin	cot	tan	cosec	sec	cos	sin	cot	tan	cosec	sec	cos	sin	cot	tan	cosec	sec	′
	65°						64°						63°						62°						

Figure 12-15 (continued)

CALCULATING SINE BAR VALUES

When using a sine bar to check an angle or to set up a workpiece for a machining operation, you will have to calculate the proper stack height and angles. The formulas are as follows.

To calculate the stack height:

$$H = \text{sine} \times L$$

To calculate the angle:

$$\text{Sine} = \frac{H}{L}$$

where
H = gage block stack height
Sine = sine of the desired angle
L = length of the sine bar

When using these formulas, you will need a trigonometric functions table (Figure 12-15) to find both the angle and the sine of the angle. For example, find the height of gage blocks needed to check the angle shown in Figure 12-16 using a 10-in. sine bar.

$$H = \text{sine} \times L$$

$$H = \text{sine } 24° \ 10' \times 10$$

$$H = 0.40939 \text{ (from Figure 12-15)} \times 10$$

$$H = 4.0939''$$

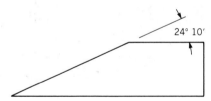

Figure 12-16 Calculating the gage block stack height.

If the stack height is known and the angle must be found, the second formula is used. For example, compute the angle of the 10-in. sine bar shown in Figure 12-17.

$$\text{Sine} = \frac{H}{L}$$

$$\text{Sine} = \frac{3.7461}{10}$$

$$\text{Sine} = 0.37461$$

$$0.37461 = 22° \text{ (from Figure 12-15)}$$

One other method you can use for calculating sine bar values is to use a table of sine bar constants. These tables have all the values directly computed and

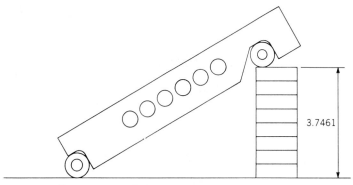

Figure 12-17 Calculating the angle of the sine bar.

listed for specific sine bar lengths, usually 5 in. If you are using a 5-in. sine bar, all you have to do is find the desired angle and read the stack height directly from the chart. If you have a 10-in. sine bar, all you have to do is find the value on the chart and multiply the value by 2.

ANGULAR GAGE BLOCKS

Angular gage blocks are the most accurate angular measuring tool commonly used in the shop. These gage blocks are available in sets with varying numbers of blocks in ranges from 0 to 99°. The three standard grades of angular gage blocks have accuracies (in seconds) of $1''$, $\frac{1}{2}''$, and $\frac{1}{4}''$.

These blocks are wrung together in the same way as regular gage blocks. The unique feature of these gage blocks is their ability to add or subtract from each other. As shown in Figure 12-18, each block has a plus (+) end and a minus (−) end. If the plus marks are aligned, as shown on the left, the values are added. If the plus and minus marks are aligned, as shown on the right, the value of the top block is subtracted from the bottom block.

Angular gage blocks are very precise measuring tools. You should handle these blocks carefully and treat them the way you would regular gage blocks.

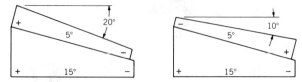

Figure 12-18 Angular gage blocks.

SELF TEST

1. Into how many parts can a degree be divided using precision angular measuring tools? (Circle one.)

    a. 60

    b. 100

    c. 1400

    d. 14,000

150 SECTION V ANGULAR MEASURING TOOLS

2. What is the name of the dimensioned angle in a triangle? (Circle one.)

   a. Opposite

   b. Complementary

   c. Working

   d. Adjacent

3. What is the name of the angle opposite the dimensioned angle in a triangle? (Circle one.)

   a. Opposite

   b. Complementary

   c. Working

   d. Adjacent

4. Find angle X for parts a to d of Figure 12-19.

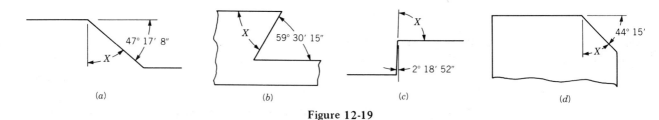

Figure 12-19

5. How accurate is the vernier bevel protractor? (Circle one.)

   a. 5′

   b. 0.5′

   c. 5″

   d. 0.5″

6. What are the values shown on the vernier bevel protractors in parts a to b of Figure 12-20?

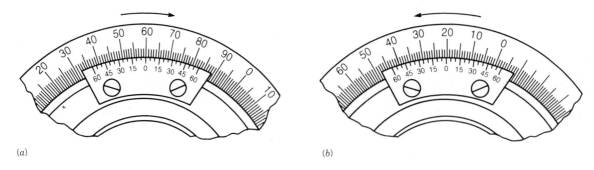

Figure 12-20

7. Which part of a triangle does the sine bar represent? (Circle one.)

    a. Opposite side

    b. Top side

    c. Adjacent side

    d. Hypotenuse

8. What part of a sine bar setup is used to represent the opposite side of a triangle? (Circle one.)

    a. Sine bar

    b. Surface plate

    c. Gage block stack

    d. Cylindrical plugs

9. What are the standard lengths of inch-type sine bars? (Circle one.)

    a. 5 and 12 in.

    b. 10 and 12 in.

    c. 12 and 6 in.

    d. 5 and 10 in.

10. What are the standard lengths of metric sine bars? (Circle one.)

    a. 100 and 200 mm

    b. 150 and 300 mm

    c. 500 and 1000 mm

    d. 1000 and 2000 mm

11. Which of the following must also be used to calculate sine bar values? (Circle one.)

    a. Table of sine bar constants

    b. Table of sine bar sizes

    c. Table of trigonometric functions

    d. Either a or c

    e. Either b or c

12. How accurate are angular gage blocks? (Circle one.)

    a. $1°, \frac{1}{2}°, \frac{1}{4}°$

    b. $1', \frac{1}{2}', \frac{1}{4}'$

    c. $1'', \frac{1}{2}'', \frac{1}{4}''$

    d. None of the above

SECTION V ANGULAR MEASURING TOOLS

13. What feature of angular gage blocks makes them unique? (Circle one.)

    a. Their ability to measure angles

    b. Their ability to add or subtract from each other

    c. Their ability to add to angular values

    d. Their ability to subtract from angular values

14. How are angular gage blocks held together? (Circle one.)

    a. Wringing

    b. Screws

    c. Magnetism

    d. Friction

15. What is the value of angle X in Figure 12-21?

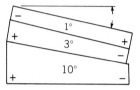
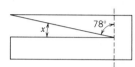

Figure 12-21

Answers to Self Test

1. d
2. c
3. b
4. a. = 42° 42′ 52″
   b. = 59° 30′ 15″
   c. = 87° 41′ 8″
   d. = 45° 45′
5. a
6. a. = 59° 45′
   b. = 20° 20′

7. d
8. c
9. d
10. a
11. d
12. c
13. b
14. a
15. 12°

Answers to Practice Problems

1. a. 38° 43′
   b. 45° 4′ 50″
   c. 63° 20′ 34″
   d. 39° 15′
   e. 46° 41′ 38″
2. a. 25°
   b. 30° 15′
   c. 35° 5′

   d. 49° 25′
   e. 35°
   f. 49° 45′
   g. 49° 15′
   h. 30°
   i. 49° 15′
   j. 55° 5′

SECTION VI — LAYOUT

UNIT 13　**Nonprecision Layout**

UNIT 14　**Precision Layout**

UNIT 13 # Nonprecision Layout

Layout is a machine shop process of transferring information from a print to a workpiece. The purpose of layout is to give the machinist an accurate, full-size description of the areas of the part that are to be machined. Layout may involve simply locating a few holes or completely drawing the entire object.

Layout can be divided into two general categories, nonprecision layout and precision layout. Nonprecision layout is generally accurate only to $\frac{1}{64}$ in.; precision layout is accurate to 0.005 in. or closer if necessary.

LAYOUT TOOLS The layout tools most commonly used for nonprecision layout are the combination square, steel rule, scribe, dividers, surface gage, prick punch, and center punch.

COMBINATION SQUARE
The combination square is used for laying out parallel lines and scribing lines at right angles (Figure 13-1). For lines that must be drawn a specific distance from an edge, the square is preset to the proper size and the line is drawn as shown in Figure 13-2. The center head attachment is used to locate the centers of cylindrical objects (Figure 13-3).

STEEL RULES
The steel rule is used to make general measurements, set other layout tools, and as a straightedge for connecting layout marks or finding centers (Figure 13-4).

SCRIBES
Scribes (Figure 13-5) are used for drawing layout lines. The two most common types of scribes are the machinists' scribe and the pocket scribe. The machinists' scribe is double ended, with one straight and one hooked end. The pocket scribe has a removable point that can be reversed in the handle. Both of these scribes are widely used for all types of layout.

156 SECTION VI LAYOUT

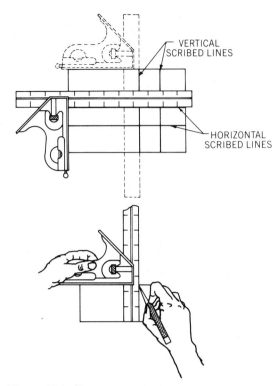

Figure 13-1 Constructing lines with a combination square.

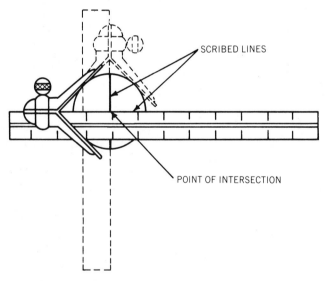

Figure 13-3 Using a center head to find the center of a cylindrical part.

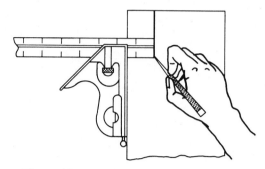

Figure 13-2 Using the end of the blade in a combination square to mark lines a specific distance from an edge.

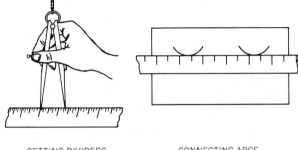

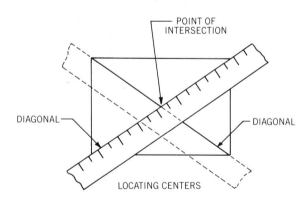

Figure 13-4 Using a steel rule for layout work.

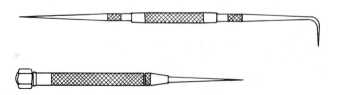

Figure 13-5 Scribes.

DIVIDERS, HERMAPHRODITE CALIPERS, AND TRAMMELS

Dividers, hermaphrodite calipers, and trammels are tools used for all types of layout operations. The divider (Figure 13-6) is used for scribing arcs and circles and marking equal distances (Figure 13-7).

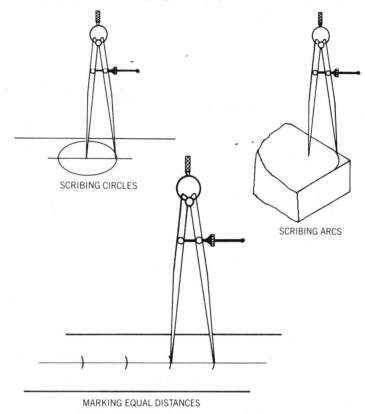

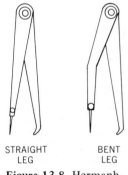

Figure 13-6 Dividers.

Figure 13-7 Using dividers.

The hermaphrodite caliper (Figure 13-8), with one divider leg and one caliper leg, is very useful for a variety of layout applications. Figure 13-9 shows several uses of this tool, including scribing lines parallel to inside and outside edges, scribing lines at right angles, and finding the centers of cylindrical workpieces. As shown in Figure 13-10, the steel rule is used to set the opening of the hermaphrodite caliper.

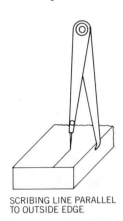

Figure 13-8 Hermaphrodite calipers.

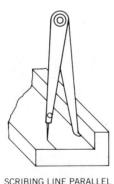

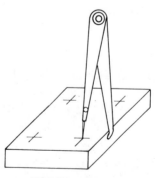

Figure 13-9 Using hermaphrodite calipers.

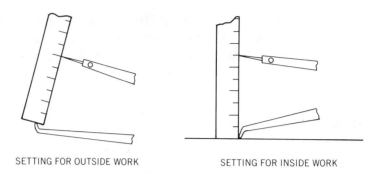

Figure 13-10 Setting hermaphrodite calipers.

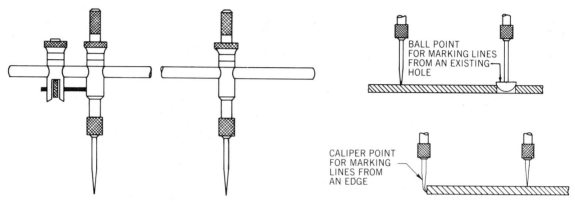

Figure 13-11 Trammels.

Figure 13-12 Using trammels.

The trammel (Figure 13-11) is generally available in two styles, the plain trammel and the beam trammel. These tools are used for layouts that are beyond the normal range of the standard dividers and hermaphrodite calipers. With attachments, they can be used for locating lines from holes or edges as well as scribing large arcs and circles (Figure 13-12).

SURFACE GAGES

The surface gage (Figure 13-13) is used for a wide variety of layout operations. Figure 13-14 shows a typical layout operation using the surface gage. Surface gages are generally used on a surface plate or similar flat surface. The combination square is generally used to set the height of the surface gage (Figure 13-15).

PUNCHES

The most common punches used for layout work are the prick punch and the center punch (Figure 13-16). The prick punch is used to make slight indentations for locating divider or trammel points; the center punch is used to deepen the prick punch marks for drilling. It is good practice when using a punch to locate the point by tilting the punch to one side. Once the point is exactly on the layout lines, the punch is brought to a vertical position and carefully tapped with a hammer (Figure 13-17). A useful variation of the center punch is the automatic center punch, shown in Figure 13-18, which has a spring-loaded striking mechanism, eliminating the need for a hammer.

Unit 13   Nonprecision Layout   159

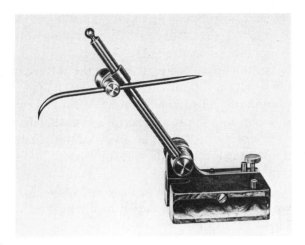

Figure 13-13 Surface gage. (Courtesy of DoAll Co.)

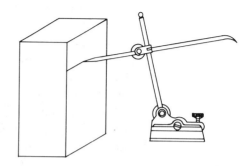

Figure 13-14 Using a surface gage.

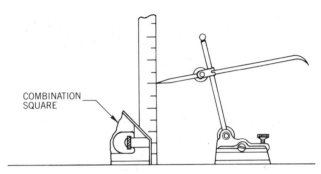

Figure 13-15 Setting a surface gage.

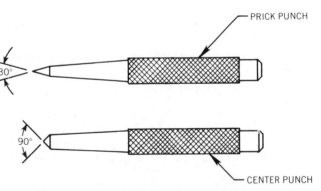

Figure 13-16 Punches used for layout.

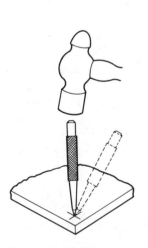

Figure 13-17 Using a punch.

Figure 13-18 Automatic center punch.

LAYOUT PROCEDURE

The first step in laying out any part is studying the print. Look for all details and dimensions that will have an affect on the layout. Once you are sure you know exactly what must be done, you can begin the layout.

To make layout lines clearly visible, apply a layout ink to the surface of the workpiece. This layout ink is generally a blue coating that is available in either liquid or aerosol form and is commonly called blueing; it dries fast and permits clear, crisp layout lines to be drawn on almost any surface. Another form of layout fluid commonly used for castings is a mixture of powdered chalk and either water or alcohol. Before applying either of these layout fluids, make sure the surface is clean and free from any grease or oil.

When the layout fluid has dried, you are ready to begin your layout. To demonstrate layout procedures, we will use the part shown in Figure 13-19. The stock size we will use is 6 × 3 in.

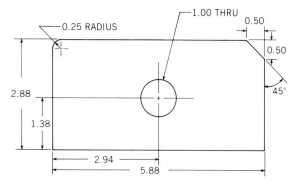

Figure 13-19 Layout problem.

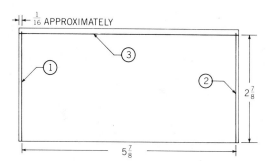

Figure 13-20 Laying out outside form.

1. Using a combination square and scribe, mark line 1 approximately $\frac{1}{16}$ in. in from the left edge (Figure 13-20).

2. Measure the length of the part with a rule and mark the part at $5\frac{7}{8}$ in. Using the combination square and scribe, mark line 2 (Figure 13-20).

3. Using a hermaphrodite caliper, set at $2\frac{7}{8}$ in., mark line 3 (Figure 13-20).

The basic rectangular outline of the part is now complete. Before going on, however, you should check all dimensions and make sure everything is correct. Now let's finish the details of the part.

4. Using a steel rule, set your dividers at $\frac{1}{4}$ in. Starting in the upper left corner, mark arcs 1 and 2. Reposition the dividers and mark arcs 3 and 4 (Figure 13-21). Using the same divider setting, mark the radius 5.

5. Next, reset your dividers to $\frac{1}{2}$ in. and, starting in the upper right corner, mark arcs 1 and 2 (Figure 13-22). Now connect the marks with a steel rule and scribe.

6. To locate the center hole, set your hermaphrodite calipers at $1\frac{3}{8}$ in. and mark line 1. Next, using a steel rule and scribe, mark a line $2\frac{15}{16}$ in. from the left vertical line along line 1. Using a combination square and scribe, mark line 2 (Figure 13-23).

Unit 13 Nonprecision Layout   161

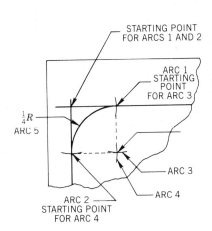

Figure 13-21 Laying out a radius.

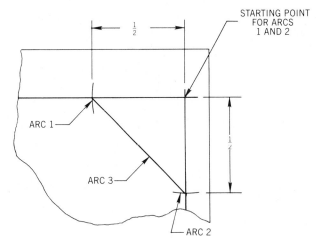

Figure 13-22 Laying out a 45° angle.

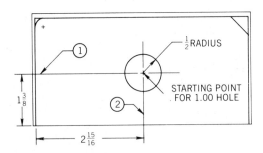

Figure 13-23 Laying out a center hole.

7. The final step is marking the 1-in. diameter hole. To do this, set your dividers at $\frac{1}{2}$ in. and mark a circle where lines 1 and 2 intersect (Figure 13-23). When marking the circle, it is good practice to prick punch the intersection of the lines to permit the divider to make an accurate circle without sliding or moving off the work.

The part is now laid out and ready to machine. Keep in mind, during the machining operation, the location of the scrap. Always cut a part on the scrap side of the layout lines. Otherwise your part will be too small.

SELF TEST

1. How accurate is nonprecision layout? (Circle one.)

   a. 0.015 in.

   b. $\frac{1}{16}$ in.

   c. 0.005 in.

   d. $\frac{1}{64}$ in.

2. How accurate is general precision layout? (Circle one.)
   a. 0.015 in.
   b. $\frac{1}{16}$ in.
   c. 0.005 in.
   d. $\frac{1}{64}$ in.

3. List two primary uses of the combination square in layout work.

4. What attachment for the combination square is generally used to find the center of cylindrical workpieces? (Circle one.)
   a. Square head
   b. Center square
   c. Protractor head
   d. Center head

5. What tool is usually used to set dividers and hermaphrodite calipers? (Circle one.)
   a. Combination square
   b. Surface gage
   c. Steel rule
   d. Shrink rule

6. What types of punches are generally used for layout work? (Circle one.)
   a. Drive and prick
   b. Prick and spot
   c. Spot and center
   d. Center and prick

7. What type of layout fluid is generally used for most layouts? (Circle one.)
   a. Blueing
   b. Red soap mixture
   c. Chalk mixture
   d. a and b
   e. a and c

8. Before beginning a layout, what should your first step be? (Circle one.)
   a. Apply layout fluid
   b. Check the print
   c. Clean the part
   d. Center punch hole locations

9. List five common layout tools.

10. What layout tool should be used to layout the part details in Figure 13-24?

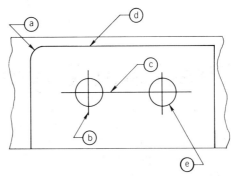

Figure 13-24

Answers To Self Test

1. d
2. c
3. a. Marking parallel lines
   b. Marking lines at right angles
4. d
5. c
6. d
7. e
8. b
9. Any five of the follwing are correct.
   a. Combination square
   b. Steel rule
   c. Scribe
   d. Dividers
   e. Hermaphrodite calipers
   f. Trammels
   g. Surface gage
   h. Prick punch
   i. Center punch
10. a. Dividers
    b. Combination square
    c. Hermaphrodite calipers
    d. Hermaphrodite calipers
    e. Dividers

UNIT 14 # Precision Layout

Precision layout is usually performed on parts that must be made to close tolerances, such as parts used in jigs and fixtures or metal stamping dies. Generally, precision layout is accurate to within 0.005 in. There are cases where even finer tolerances are required, but these cases are rare.

TOOLS USED FOR PRECISION LAYOUT

The primary tools used for precision layout are the vernier height gage, surface plate, gage blocks and accessories, and an assortment of various-size angle plates and vee blocks.

VERNIER HEIGHT GAGE

The vernier height gage (Figure 14-1) is used to mark horizontal lines to an accuracy of 0.001 in. In cases where intersecting lines are required, the part is simply rotated 90° (Figure 14-2). Vernier height gages are set by determining the vertical distance from the bottom edge of the part to each horizontal location. This generally involves setting the vernier to the first horizontal line and adding the distance to the next line to the value shown on the vernier scales. For example, if the first line were dimensioned as 1.00 in. from the edge and the next line was 1.25 in. from the first, the vernier would be set at 1.000 in. for the first line and at 2.250 in. for the second. Vernier height gages are read in the same way as vernier calipers and are equipped with either 25 or 50 division vernier scales.

Vernier height gages are equipped with a removable scribing point (Figure 14-3). This point must always be used when laying out any part with the height gage. Never use the slide arm for layout. This could cause damage to the slide arm and affect the accuracy of the height gage. Always use the scribing point.

SURFACE PLATES

The surface plate (Figure 14-4) is generally used for all precision layout operations. These plates are made of cast iron, semisteel, or granite. Surface plates are flat to within 0.0001 in. and have a very smooth working surface. When using a surface plate, be careful not to chip or scratch the working surface. Never drop anything on a surface plate and never drag heavy parts across the working surface. If the part you are laying out is heavy, get help. Whenever

Unit 14 Precision Layout 165

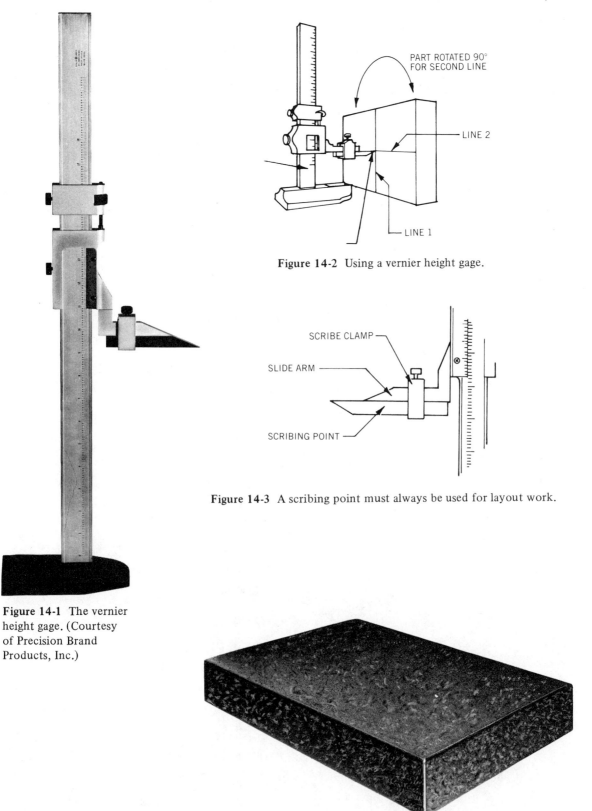

Figure 14-1 The vernier height gage. (Courtesy of Precision Brand Products, Inc.)

Figure 14-2 Using a vernier height gage.

Figure 14-3 A scribing point must always be used for layout work.

Figure 14-4 Surface plate. (Courtesy of Precision Brand Products, Inc.)

possible, roll the part onto the surface plate instead of setting it down. Remember that the surface plate is an accurate tool. Treat it accordingly.

GAGE BLOCKS AND ACCESSORIES

Gage blocks and their accessories (Figure 14-5) are used for a variety of precision layout applications. Gage blocks can be used to space part details in much the same way as a divider, or as shown in Figure 14-6, as a height gage to mark horizontal lines accurately. When gage blocks are used for layout, a surface plate must be used for maximum accuracy. When using gage blocks, keep the surface plate clean to avoid scratching the base block.

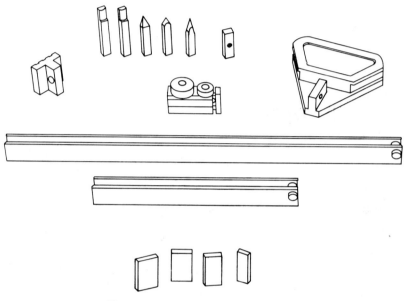

Figure 14-5 Gage block accessories.

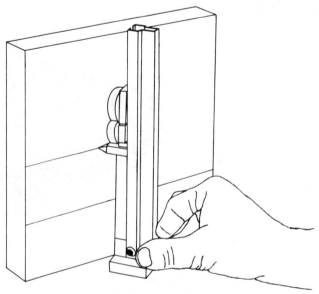

Figure 14-6 Using gage blocks as a height gage for layout.

ANGLE PLATES AND VEE BLOCKS

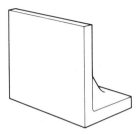

Figure 14-7 Angle plate.

Angle plates (Figure 14-7) are generally used to hold and reference workpieces for layout. As shown in Figure 14-8, workpieces are held against the angle plate by toolmakers' parallel clamps (Figure 14-9). Angle plates generally have all sides ground to precise accuracies. The angle plate can be used flat or set up on either side for laying out intersecting lines. To insure accuracy when using an angle plate, clean both the angle plate and the workpiece before mounting the workpiece. When clamping the part, try to visualize every step of the layout so you can position the toolmakers' clamps in a location where they will not interfere with any layout operations.

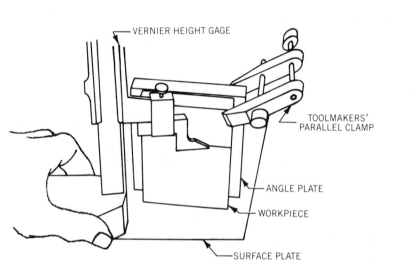

Figure 14-8 Using an angle plate for layout.

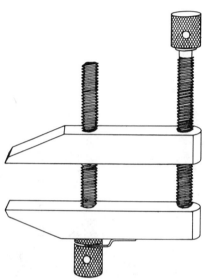

Figure 14-9 Toolmakers' parallel clamp.

Vee blocks (Figure 14-10) are used to locate and support cylindrical workpieces accurately for layout. As shown, vee blocks have two vee grooves. Each vee groove is accurately ground parallel to the base and at exactly 90°. The larger vee is used for larger diameters, and the smaller vee is used for smaller diameters. Most vee blocks are equipped with a clamp that holds the workpiece in the vee block. Vee blocks are generally used in pairs of matched blocks to permit accurate location of parts that would normally be too long for a single block. Figure 14-11 shows how a vee block can be used with a vernier height gage to locate the center of a workpiece accurately.

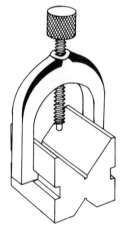

Figure 14-10 Vee block and clamp.

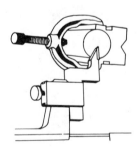

Figure 14-11 Using a vee block for layout.

SELF TEST

1. What is the principal difference between precision and nonprecision layout?

2. What must always be used with a vernier height gage for layout work?

3. How accurate are vernier height gages? (Circle one.)

   a. $\frac{1}{32}$ in.

   b. $\frac{1}{64}$ in.

   c. 0.001 in.

   d. 0.0001 in.

4. Which of the following is not a material generally used for surface plates? (Circle one.)

   a. Semisteel

   b. Steel plate

   c. Granite

   d. Cast iron

5. How should you put a workpiece on a surface plate? (Circle one.)

   a. No particular method

   b. Gently set it down

   c. Roll it onto the plate

   d. None of the above

6. Which of the following is a typical use for gage blocks in layout? (Circle one.)

   a. With a base block, as a height gage

   b. As a divider for very small holes

   c. As a hermaphrodite caliper

   d. To measure proper spacings

7. What is the purpose of an angle plate in layout?

8. How are workpieces usually held on an angle plate?

9. What is the angle of the grooves in a vee block? (Circle one.)

   a. 45°

   b. 60°

   c. 90°

   d. 100°

10. Why are vee blocks made in pairs?

Answers To Self Test

1. The discrimination of the tools used to mark the lines
2. The scribing point
3. c
4. b
5. c
6. a
7. To hold and reference workpieces for layout
8. With toolmakers' parallel clamps
9. c
10. So they can be used to hold parts that are too long for one vee block

APPENDIXES

APPENDIX A

Glossary of New Words

Accuracy	The degree of conformity to an established standard.
Angular Measurement	Measurement of angles. The measurement units normally used with angular measurements are degrees.
Comparison Measurement	Measurement made with instruments that compare the unknown size to a known value, for example, a dial indicator.
Decimal Inch	An inch unit that has its lesser units described in decimal units.
Dimension	The exact size or locational value specified on a shop print.
Direct Measurement	Measurement made directly with a measuring tool.
Discrimination	The finest division of a measuring tool that can be read reliably.
Error of Bias	Error caused by either intentional or unintentional bias in measurement. In other words, if you want the micrometer to read a certain size, you may overtighten or overtwist the tool to make the desired reading.
Feel	The amount of drag between the measuring tool and the workpiece as felt by the hands of the measurer.
Fractional Inch	An inch unit that has its lesser units described in fractional units.
Gages	Tools used to check the conformity of a part feature to a standard of specific size or shape.
Graduations	The equally spaced lines on each edge of a rule, or the sleeve and thimble of a micrometer. These lines are used to denote units of measurement.
Indirect Measurement	Measurement made with the aid of transfer measuring tools in conjunction with direct measuring tools.
Instrument Error	Error caused by either poor quality tools or defective measuring tools. Instrument error can also be caused by trying to use an instrument to a discrimination finer than it is designed to read.
Limits	The maximum and minimum size of a part feature as determined by the tolerance.
Line of Measurement	An imaginary straight line between the reference point and the measured point.
Linear Measurement	Measurement of length, or straight line distances. The measurement units normally used with linear measurements are the inch and millimeter.
Manipulative Error	Error caused by misreading or mishandling the measuring tool due to the way the measuring tool was referenced to the work.
Measured Point	The line or edge that terminates a measurement. The end of a measurement.

Observational Error	Error caused by improper reading due to poor visual alignment with the measured point.
Pitch	The distance from a point on one thread to the same point on the next adjacent thread.
Plug Gages	Gages that are intended to check the diameter or shape of a round hole.
Precision	The repeatability of a measuring process.
Range	The capacity of the measuring tool.
Reference Point	The base from which a measurement is taken. The beginning of a measurement.
Reliability	The condition where the actual results are the same as the predicted or desired results.
Scales	The series of lines on each edge of a rule.
Stack Height	The height of the gage blocks needed to provide a desired angle with a sine bar.
Standard	An established and known value used to measure an unknown quantity.
Taper Plug Gages	Gages that are intended to check the size or shape of a tapered hole.
Threaded Plug Gages	Gages that are intended to check the pitch diameter of threaded holes.
Tolerance	The total amount of permitted variation from the basic size dimension of a part feature.
Vernier Scales	A system of measurement that uses sliding scales to make measurements.

APPENDIX B
Common Conversions

When You Know	Multiply By	To Find
Cubic inches	0.01639	Liters
Degrees	3600	Seconds
Diameter	3.1416	Circumference
Diameter	0.7854 diameter^2	Area
Feet	0.3048	Meters
Feet	304.8	Millimeters
Feet/minute	0.01136	Miles/hour
Gallons	3.785	Liters
Grams	0.03527	Ounces
Horsepower	33,000	Foot-pound/minute
Horsepower	745.7	Watts
Inches	25.4	Millimeters
Inches	0.0278	Yards
Kilograms	2.205	Pounds
Kilometers	3,281	Feet
Kilometers	0.6214	Miles
Kilometers	1,094	Yards
Kilometers/hour	0.6214	Miles/hour
Kilowatts	1.341	Horsepower
Knots	1.152	Miles/hour
Miles	5,280	Feet
Miles	1.609	Kilometers
Miles	1,760	Yards
Miles/hour	88	Feet/second
Miles/hour	0.8684	Knots
Minutes	0.01667	Degrees
Liters	61.02	Cubic inches
Liters	0.2642	Gallons
Meters	3.281	Feet
Meters	39.37	Inches
Meters	1.094	Yards
Pounds	0.4536	Kilograms
Square feet	0.0929	Square meters
Square inches	645.2	Square feet
Yards	0.9144	Meters

APPENDIX C   # Temperature Conversion

HOW TO USE THE TEMPERATURE CONVERSION CHART

The chart is arranged in three columns. The degree (°) or first column is whatever scale you wish to convert from. It can be Fahrenheit degrees *or* Celsius degrees. The second column, F, is Fahrenheit degrees. The third column, C, is Celsius degrees.

SAMPLE

Convert 68° Celsius to Fahrenheit degrees. Find the ° or degree column. Look down the column until you find 68°. Look on the same line under F. You will see there that 68° Celsius converts to 154.4° Fahrenheit.

Convert 68° Fahrenheit to Celsius degrees. Find the ° or degree column. Look down the column until you find 68°. Look on the same line under C. You will see there that 68° Fahrenheit converts to 20.0° Celsius.

There is no relationship between the F and C columns, only between the degree (°) column and *one* of the other (F or C) columns.

TEMPERATURE CONVERSION CHART

°	F	C	°	F	C	°	F	C	°	F	C
-10	14	-23	29	84.2	- 1.7	59	138.2	15.0	89	192.2	31.7
0	32	-17.8	30	86.0	- 1.1	60	140.0	15.6	90	194.0	32.2
1	33.8	-17.2	31	87.8	- 0.6	61	141.8	16.1	91	195.8	32.8
2	35.6	-16.7	32	89.6	0.0	62	143.6	16.7	92	197.6	33.3
3	37.4	-16.1	33	91.4	0.6	63	145.4	17.2	93	199.4	33.9
4	39.2	-15.6	34	93.2	1.1	64	147.2	17.8	94	201.2	34.4
5	41.0	-15.0	35	95.0	1.7	65	149.0	18.3	95	203.0	35.0
6	42.8	-14.4	36	96.8	2.2	66	150.8	18.9	96	204.8	35.6
7	44.6	-13.9	37	98.6	2.8	67	152.6	19.4	97	206.6	36.1
8	46.4	-13.3	38	100.4	3.3	68	154.4	20.0	98	208.4	36.7
9	48.2	-12.8	39	102.2	3.9	69	156.2	20.6	99	210.2	37.2
10	50.0	-12.2	40	104.0	4.4	70	158.0	21.1	100	212.0	37.8
11	51.8	-11.7	41	105.8	5.0	71	159.8	21.7	110	230	43
12	53.6	-11.1	42	107.6	5.6	72	161.6	22.2	120	248	49
13	55.4	-10.6	43	109.4	6.1	73	163.4	22.8	130	266	54
14	57.2	-10.0	44	111.2	6.7	74	165.2	23.3	140	284	60
15	59.0	- 9.4	45	113.0	7.2	75	167.0	23.9	150	302	66
16	60.8	- 8.9	46	114.3	7.8	76	168.8	24.4	160	320	71
17	62.6	- 8.3	47	116.6	8.3	77	170.6	25.0	170	338	77
18	64.4	- 7.8	48	118.4	8.9	78	172.4	25.6	180	356	82
19	66.2	- 7.2	49	120.2	9.4	79	174.3	26.1	190	374	88
20	68.0	- 6.7	50	122.0	10.0	80	176.0	26.7	200	392	93
21	69.8	- 6.1	51	123.8	10.6	81	177.8	27.2	210	419	99
22	71.6	- 5.6	52	125.6	11.1	82	179.6	27.8	212	413.6	100
23	73.4	- 5.0	53	127.4	11.7	83	181.4	28.3	220	428	104
24	75.2	- 4.4	54	129.2	12.2	84	183.2	28.9	230	446	110
25	77.0	- 3.9	55	131.0	12.8	85	185.0	28.4	240	464	116
26	78.8	- 3.3	56	132.8	13.3	86	186.8	30.0	250	482	121
27	80.6	- 2.8	57	134.6	13.9	87	188.6	30.6	260	500	127
28	82.4	- 2.3	58	136.4	13.4	88	190.4	31.1	270	518	132

APPENDIX D Standard Metric and American Threads

STANDARD METRIC SCREW THREADS

Nominal Size (mm)	Pitch P (mm)	Basic Thread Designation *	Nominal Size (mm)	Pitch P (mm)	Basic Thread Designation *
1.6	0.35	M1.6	14	2	M14
1.8	0.35	M1.8		1.5	M14 × 1.5
2	0.4	M2	16	2	M16
2.2	0.45	M2.2		1.5	M16 × 1.5
2.5	0.45	M2.5	18	2.5	M18
3	0.5	M3		1.5	M18 × 1.5
3.5	0.6	M3.5	20	2.5	M20
4	0.7	M4		1.5	M20 × 1.5
4.5	0.75	M4.5	22	2.5	M22
5	0.8	M5		1.5	M22 × 1.5
6	1	M6	24	3	M24
7	1	M7		2	M24 × 2
8	1.25	M8	27	3	M27
	1	M8 × 1		2	M27 × 2
10	1.5	M10	30	3.5	M30
	1.25	M10 × 1.25		2	M30 × 2
12	1.75	M12	33	3.5	M33
	1.25	M12 × 1.25		2	M33 × 2
			36	4	M36
				3	M36 × 3
			39	4	M39
				3	M39 × 3

TAP DRILL SIZES FOR METRIC THREADS

Thread	Tap Drill	Thread	Tap Drill	Thread	Tap Drill	Thread	Tap Drill
M1.6	1.25	M6	5.00	M16	14.00	M27	24.00
M1.8	1.45	M7	6.00	M16 × 1.5	14.50	M27 × 2	25.00
M2	1.60	M8	6.70	M18	15.50	M30	26.50
M2.2	1.75	M8 × 1	7.00	M18 × 1.5	16.50	M30 × 2	28.00
M2.5	2.05	M10	8.50	M20	17.50	M33	29.50
M3	2.50	M10 × 1.25	8.70	M20 × 1.5	18.50	M33 × 2	31.00
M3.5	2.90	M12	10.20	M22	19.50	M36	32.00
M4	3.30	M12 × 1.25	10.80	M22 × 1.5	20.50	M36 × 3	33.00
M4.5	3.70	M14	12.00	M24	21.00	M39	35.00
M5	4.20	M14 × 1.5	12.50	M24 × 2	22.00	M39 × 3	36.00

TAP DRILL SIZES UNC-UNF INCH SCREW THREADS

Thread	Tap Drill	Decimal Equivalent	Major Diameter Screw	Thread	Tap Drill	Decimal Equivalent	Major Diameter Screw
2–56	50	0.070	0.086	$\frac{3}{8}$–16	O	0.316	0.375
2–64	49	0.073		$\frac{3}{8}$–24	Q	0.332	
4–40	43	0.089	0.112	$\frac{7}{16}$–14	U	0.368	0.4375
4–48	42	0.0935		$\frac{7}{16}$–20	W	0.386	
6–32	36	0.1065	0.138	$\frac{1}{2}$–13	$\frac{27}{64}$	0.4219	0.500
6–40	33	0.113		$\frac{1}{2}$–20	$\frac{29}{64}$	0.4531	
8–32	29	0.136	0.164	$\frac{9}{16}$–12	$\frac{31}{64}$	0.4884	0.5625
8–36	28	0.1405		$\frac{9}{16}$–18	$\frac{33}{64}$	0.5156	
10–24	25	0.1495	0.190	$\frac{5}{8}$–11	$\frac{17}{32}$	0.5133	0.625
10–32	21	0.159		$\frac{5}{8}$–18	$\frac{37}{64}$	0.5781	
12–24	16	0.177	0.216	$\frac{3}{4}$–10	$\frac{21}{32}$	0.6563	0.750
12–28	14	0.182		$\frac{3}{4}$–16	$\frac{11}{16}$	0.6875	
$\frac{1}{4}$–20	7	0.201	0.250	$\frac{7}{8}$– 9	$\frac{49}{64}$	0.7656	0.875
$\frac{1}{4}$–28	3	0.213		$\frac{7}{8}$–14	$\frac{13}{16}$	0.8125	
$\frac{5}{16}$–18	F	0.257	0.3125	1 – 8	$\frac{7}{8}$	0.875	1.000
$\frac{5}{16}$–24	I	0.272		1 –12	$\frac{59}{64}$	0.9219	

Index

Accuracy, 10
Adjustable squares, 32-36, 122
   variations on, 36-37
Angle gages, 110
Angle plates, 167
Angular gage blocks, 149
Angular measurement, 129-130, 140. *See also* Angular measuring tools
Angular measuring tools, 127-149
   nonprecision, 129-136
   precision, 140-149
Automatic center punch, 158

Back plunger indicator, 90
Bevel protractors, 132-134
   vernier, 141-144
Bevels, 136
Bore gages, dial, 92

Caliper, dial, 69-72
Center gages, 114
Center punch, 158
Combination squares, 32-34, 155, 160
Comparison gages, 109-116
   angle gages, 110
   center gages, 114
   drill gages, 109
   edge finder, 112
   form gages, 114
   radius gages, 115
   screw pitch gages, 114
   sheet, plate, and wire gages, 109
   as size gages, 109-116
   taper gages, 111-112
   thickness gages, 110
   toolmakers' buttons, 113-114
   29° screw thread gages, 115
   wigglers, 113
Conversions, 7, 173-175
Cross test levels, 120
Cubits, 4
Cylindrical squares, 122-123

Depth gage, 36
   combination protractor and, 135-136
   dial, 90
Dial bore gages, 92
Dial caliper, 69-72
Dial depth gage, 90
Dial indicators, 81-93
   attachments for mounting, 86

   balanced, 82
   construction of, 81-82
   continuous, 82
   metric, 82, 85
   reading, 82-85
   revolution counter, 82
   using, 86-88
   variations of, 89-93
   zeroing dial, 86-88
Dial snap gage, 92
Dial thickness gage, 92
Diemakers' squares, 35-36
Dimensions, 12
Discrimination of scale, 21
Dividers, 157, 160, 161
Double squares, 34-35
Drill gages, 109
Drill point gages, 39

Edge finder, 112
English system of linear measurement, 5-6
Errors of bias, 27

Factors that affect measurement, 12
Feel and transfer measurements, 27, 28, 57
Fixed gages, 103-106
   care of, 106
   GO-NO GO method of gaging, 103-104
   plug gages, 104-105
   ring gages, 105-106
   snap gages, 106
Ford, Gerald, 5
Form gages, 114
Function of measurement, 3

Gage blocks, 98-103, 166
   accessories, 102-103, 166
   accuracy of, 99-100
   angular, 149
   calculating, stack height, 101-102
   care of, 103
   types of, 98-99
   using, 100
   wringing, 100
Glossary, 171-172
GO-NO GO method of gaging, 103-104
Graduated measuring tools, 15-39
   adjustable squares and rule attachments, 32-39

   steel rules, 17-30

Hermaphrodite calipers, 157, 160
History of measurement, 3-5

Indicators, dial, *see* Dial indicators
Instrument error, 24
International System of Units (SI), 6-7, 23

Keyseat clamps, 38

Language of measurement, 9-13
Layout, 153-167
   nonprecision, 155-161
   precision, 164-167
Levels, 119-121
   cross test, 120
   machinists', 119
   pocket, 119
   precision, 120-121
Limits, 12
Line of measurement, 9-10

Machinists' levels, 119
Manipulative error, 25
Measured point, 8
Measurement factors, 8-10
Measurement terms, 10-12
Metric system, 5, 6
Micrometers, 43-57
   care of, 57
   construction of, 43
   depth, 53-54
   direct measurements with, 54-55
   how it works, 43-44
   inside, 52
   making measurements with, 54-57
   outside, 51
   ratchet stop, 43
   reading:
      to 0.001 in., 45-46
      to 0.0001 in., 46-48
      to 0.01 mm, 49
      to 0.002 mm, 49-50
   thimble, 43, 44, 45, 47
   transfer measurements with, 55-57
   types of, 51-54
   vernier scale, 46-47, 49

Nonprecision angular measuring tools, 129-136

Index

angle transfer tools, 136
angular measurement, explanation of, 129-130
bevels, 136
combination protractor and depth gage, 135-136
protractors, 132-135
Nonprecision layout, 155-161
layout procedure, 160-161
layout tools, 155-159

Observational error, 25

Parallax error, 9, 25
Plug gages, 104-105
Pocket levels, 119
Precision, 10
Precision angular measuring tools, 140-149
angular gage blocks, 149
calculating angular measurements, 140
sine bar, 144-149
vernier bevel contractor, 141-144
Precision layout, 164-167
Precision levels, 120-121
Precision measuring tools, 41-76
angular, *see* Precision angular measuring tools
micrometers, 43-57
squares, *see* Precision squares
verniers, *see* Verniers
Precision squares, 121-123
adjustable, 122
cylindrical, 122-123
solid, 122
Prick punch, 158, 161
Protractor head, 136
Protractors, 132-135
bevel, 132-134
combination depth gage and, 135-136
steel, 134-135
vernier bevel, 141-144
Punches, 158

Quick reading reference numbers, 21

Radius gages, 115
Reference point, 8, 25
Reliability, 10-12
Right angle rule clamps, 39
Ring gages, 105-106
Rule clamps, 38
right angle, 39

Screw pitch gages, 114
Scribes, 155, 160
Sheet, plate, and wire gages, 109
Sine bar, 144-149
calculating, values, 148-149
trigonometric functions table, 146-147
Size gages, 109-116
Slide calipers, 36-37
Small hole gages, 55
Snap gages, 106
dial, 92
Solid squares, 122
Squares, adjustable, and rule attachments, 32-39, 122
Standard of measurement, 8
Standard metric and American threads, 176-177
Steel protractors, 134-135
Steel rules, 17-30, 155, 160
adjustable squares and rule attachments, 32-39
care of, 29-30
decimal inch scale, 20, 22-23
flexible, 18
fractional scale, 20, 21-22
gradations of, 20-23
hook, 18
making direct measurements with, 24-27
making indirect measurements with, 27-28
millimeter scale, 20, 23
narrow, 17
reading, 20-23
rule sets, 19
scale of, 20-23
spring tempered, 17
Straightedges, 121

Surface gages, 158
Surface plates, 164-166

Taper gages, 111-112
Telescoping gages, 55-57
Temperature conversions, 174-175
Test indicators, 89-90
Thickness gages, 110
dial, 92
Tolerance, 12
Toolmakers' buttons, 113-114
Trammels, 158
Transfer (indirect) measurements, 27-28, 55-57, 136
29° screw thread gages, 115

U.S. National Bureau of Standards, 100
Units of measurement, 5-7
converting, 7
English system, 5-6
International system (SI), 6-7, 23

Vee blocks, 167
Vernier bevel contractor, 141-144
reading, 142
using, 143-144
Vernier caliper, 73, 74
using, 74-76
Vernier depth gage, 73, 74
Vernier gear tooth caliper, 73, 74
Vernier height gage, 73, 74, 164
Verniers, 62-76, 141-144, 164
construction of, 62-63
reading dial-type measuring tools, 68-72
reading millimeter, 67
reading 25-division, 63-65
reading 50-division, 66
types of, 73-74
using vernier caliper, 74-76

Wigglers, 113